MADRAS FISHERIES DEPARTMENT

Bulletin, Vol. XVI.

THE INDIAN PEARL FISHERIES OF THE GULF OF MANNAR AND PALK BAY

BY

JAMES HORNELL, F.L.S., F.R.A.I.
Director of Fisheries, Madras

MADRAS
PRINTED BY THE SUPERINTENDENT, GOVERNMENT·PRESS

1922

TABLE OF CONTENTS.

THE INDIAN PEARL FISHERIES OF THE GULF OF MANNAR AND PALK BAY

BY

JAMES HORNELL, F L.S , F R A I ,
Director of Fisheries, Madras

I.—INTRODUCTORY (1922).

MY original report to the Government of Madras, made in 1905, was never published though a limited number of copies, long since exhausted, were printed. As it contained a large body of information on the history and conduct of the local pearl fisheries, nowhere else brought together, Government have been pleased to sanction its publication in a revised form as a Bulletin of the Fisheries Department. Since the inspection which this report chronicles, my acquaintance with the Indian pearl banks has increased immensely, for in 1908 I joined the Fisheries Service of Madras, and was shortly after appointed Superintendent of Pearl and Chank Fisheries in addition to my marine biological duties More intimate local knowledge has enabled me to make some useful additions to the historical notes, whilst in respect of the recommendations made for the improvement of the banks by cultural means, I have come to the conclusion that local conditions make this work too onerous and uncertain to be worth the heavy expenditure that would be involved. Otherwise the conclusions and recommendations have stood the test of experience and I find surprisingly little that requires alteration or amendment. The majority of the recommendations have been carried out and the best advice I can give my successors is to place reliance first and last on efficient inspection Experience shows that culture in any form on the banks is virtually impracticable owing to the impossibility of protecting the spat under culture from the ravages of fish Therefore as a successful natural spat-fall is erratic and its location impossible to predicate, the watchword of the officer in charge should be inspection, efficient and comprehensive But great care must be exercised to carry this work out economically, for with fisheries few and far between, expenses of inspection mount up rapidly and then wipe out the profits obtained from the fisheries leaving Government and the tax-payer not one pie the better. Governor Van Imhoff long ago sagely queried "Whether there is not more of glitter than of gold about the Pearl Fishery." My own opinion is that the attention of the officer in charge will be most profitably employed if concentrated upon the improvement and development of the always reliable and profitable chank fishery, and that work in regard to the inspection of the pearl banks should be subordinated to and dovetailed with this in order to avoid unproductive expenditure.

It has been found unnecessary to amend the original sections except in detail. Two new sections have been added as this appears necessary to make the report self-contained. One of these deals with the anatomy and development of the pearl-oyster, the other is a concise statement of the conclusions on pearl-formation in our local species to which I have come after twenty years' study of this difficult question. On the other hand the long summary of inspection results from 1885 to 1903 has been omitted as no longer of utility. Similarly a number of the inspection diagrams and plans have been dispensed with.

The most important occurrence in the history of the Indian pearl banks since 1905 was my discovery in 1913 of a bed of pearl oysters off Tondi, on the west coast of Palk Bay. The fishery that ensued in 1914 was the first ever known in this locality and though it brought little profit to Government, it greatly enlarged our knowledge of the habits of our local species, and cast light upon the hitherto obscure problem of where the beds are which replenish the banks in the Gulf of Mannar after periods when these are absolutely bare of oysters.

Much attention has also been given since 1908 to an investigation of the surface drift of the Gulf of Mannar. The results are in the main negative in that they show the drift to be very largely influenced by the monsoon winds. This means that it exhibits no reliable regularity; instead it varies with the intensity of each individual monsoon and we are no nearer to a knowledge of the factors affecting spat-fall, than that expressed by the statement that the conditions needful to induce a spat-fall depend upon a complex interaction of currents of exceptional character at the exact period when the free-swimming spat abound. These conditions are most likely to occur when eddies happen to form through the meeting of opposing currents as may happen during the intermonsoon periods of variable winds and currents   But this does not help us far forward, and so far as I can see it is impossible to differentiate in advance a season favourable for a heavy spat-fall, from one which is not

It will be noticed that one of the principal recommendations made was for the creation of a Fisheries Department under scientific control and the transfer thereto of the departmental working of the pearl and chank fisheries   Under the enlightened policy of the Government this eventually was the course pursued and it is a great gratification to me that I have been enabled to carry out the greater part of the recommendations I originally made, and to know that these have been of signal financial benefit to the Presidency.

MADRAS,
1st January 1922.   JAMES HORNELL.

II.—PREFACE TO THE ORIGINAL (1905) REPORT.

The accompanying report upon the present condition and future prospects of the pearl banks off the coast of Madura and Tinnevelly is the outcome of a request for the loan of my services made by the Government of Madras to that of Ceylon in the spring of 1904.

I appreciate most highly the honour thus done me and I have endeavoured to the utmost of my ability to discharge satisfactorily the duty laid upon me. The working up of the material, however, proved unexpectedly tedious and arduous; the volume of the material to be digested was very much greater than I had anticipated; wide historical inquiries had to be instituted and all this to be carried on concurrently with the exacting duties of my first year in office as Inspector of Pearl Banks and Marine Biologist to the Government of Ceylon. Critics will therefore, I trust, deal gently with the many shortcomings which I am conscious mar the present report; they will kindly bear in mind that it has been built up largely in fragments of time snatched from an all too scanty leisure. I should like indeed to devote further study to the inquiry, but under present circumstances I see no prospect of the necessary time-opportunity. I reluctantly decide that it is better to send in the report as it stands than to postpone its issue indefinitely.

Possibly I have striven to do more than was expected from me, but having put my hand to the work I could not refrain from the attempt to review the position in all its bearings. However many shortcomings pertain to this effort I have cleared some stumbling blocks from the way and have indicated the lines on which success may be attained eventually by the workers who will follow me.

My aim has been to sift the whole of the evidence available, historical, zoological and physiographical; to present the conclusions in a simple and succinct form and to formulate remedial measures on a practical and business-like basis.

It has not been found necessary to treat herein of the anatomy and habits of the pearl oyster as such are dealt with fully by Professor Herdman* and myself in the " Report on the Pearl Oyster

---

* Now Sir W. A. Herdman, Kt., F R S &c.

Fisheries of the Gulf of Manaar," recently published by the Royal Society

My grateful acknowledgments are made to my predecessors in this investigation; my work has been immensely facilitated by the extensive data recorded and ably presented in Mr H Sullivan Thomas' "Report on Pearl Fisheries" published by the Madras Government in 1884, and in the valuable reports made from time to time by Mr. E Thurston in the bulletins of the Madras Museum. It has always appeared to me a thousand pities that Mr. Thurston's abilities have not been utilized to a much greater extent than they have been in pearl fishery investigation, seeing how highly qualified he has proved himself for the task

I am also pleased to have this opportunity to thank Captain Carlyon, the Port Officer of Tuticorin and Superintendent of Pearl Banks, and Don Gabriel de Cruz Lazarus Motha Vaz, the Jathi Talaivan of the Paravas, for their courtesy and help during the actual work of inspection in May 1904   The Jathi Talaivan, indeed, spared no pains to put me in possession of all historical and traditional information of which he is the depositary owing to the close connexion which has existed for centuries between his fore-fathers and the pearl fishery organization; such has been of great value to me.

THE PEARL BANKS, CEYLON,
    28th March 1905   JAMES HORNELL.

III—HISTORICAL SURVEY OF THE PEARL FISHERIES OFF THE MADURA COAST.*

The well-known lack of the historical faculty among the peoples of India prior to the advent of Muhammadanism involves us in all but utter ignorance of the exact localities, course, and conduct of the pearl fisheries of the Gulf of Mannar, as well on the Ceylon as on the Indian side, until the date when European control began.

(a) THE PEARL BANKS PRIOR TO THE ARRIVAL OF THE PORTUGUESE

Anterior to the Portuguese seizure of the fisheries in 1524†, the glimpses we catch are hazy and unsatisfactory—glimpses recorded on their return home by sailors and travelling merchants belonging to other countries. Greeks, Egyptians, Barbary Moors, Arabs, Venetians and Genoese have all referred incidentally to these fisheries as among the notable sights seen during their journeyings, whereas Tamil and Sinhalese writers had no thoughts save for the glory and exploits of their kings and the advancement and excellence of their religious systems. The latter refer to pearls solely to utilize the idea of their beauty and the mystery of their origin for the purposes of their exuberant and florid imagery and in the exaggerated descriptions of the riches of their kings and temples.

When the pearl fisheries of the Gulf of Mannar were first exploited we have no hint; even two thousand years ago they were celebrated throughout the known world from China to the Mediterranean. In Rome, in the days of Pliny, pearls from the Gulf of Mannar were valued at a high price and Pliny refers to this fishery as the most productive of pearls of all parts of the world, while more than six centuries before (550—540 B.C.), Vijaya, the Aryan Conqueror of Ceylon, is said to have included rich offerings of pearls among the presents to his father-in-law, the Pandyan King of Madura. ‡

The earliest definite reference to a particular locality in the Gulf of Mannar where a fishery was carried on, occurs in the

---

* The term " Madura Coast" is employed in these pages, unless it be specified otherwise, in its wider and more ancient sense ; it signifies here the seaboard of the ancient Kingdom of Madura and therefore includes the shore of the modern district of Tinnevelly as well as that of the Rāmnād district.

† Gaspar Correa, " Lendas da India," volume II

‡ Vide the " Mahawansa "

" Periplus of the Erythræan Sea," written about the end of the first century A.D. by an Alexandrian Greek. In a description of the ports on the Indian Coast, he writes -

"Upon leaving Bakarē we come to the Ruddy Mountain "towards the south; the country which succeeds is under the "Government of Pandian; it is called Paralia* and lies almost "directly north and south; it reaches to Kolkhoi in the vicinity "of the pearl fishery. But the first port after leaving the Ruddy "Mountain is Balita and next to that is Komar which has a good "harbour and a village by the shore. . . . From Komar the "district extends to Kolkhoi and the pearl fishery which is "conducted by slaves or criminals condemned to the service; the "whole southern point of the continent is part of Pandian's "dominion. Beyond Kolkhoi there succeeds another district called "the 'Coast country,' lying along a gulf (Palk Bay) and has a "region inland called Argaru † Here and here only the pearls "obtained in the fishery at the island of Epiodorus ‡ are per-"forated and prepared for market and from the same place are "procured the muslins called Argaritic."

Ptolemy (who died in A.D. 163) adds further interesting references, mentioning in his description of the provinces, towns, and rivers of the East Coast—

"Country of the Kareoi ; in the Kolkhic Gulf, where there is "the pearl fishery, Sōsikourai and Kolkhoi, an emporium at the "mouth of the river Solen."

So unchanging are names and peoples in this district that at the present day the majority of these names can be readily recognized—in itself also a tribute to the accuracy of the two Græco-Egyptian geographers of 1,800 years ago Komar is obviously Kumari, anglicized into Comorin the Kareoi are the caste of Karaiyar or coast people—fishermen and boatmen—of whom the Parathavar or Paravas are perhaps a branch or section, described in the Tamil poem *Maturaik-kanchi* (Ll 140—144) § as men who dived for pearls or for conch shells and knew the charm to keep off sharks from that part of the sea where they dived.

---

* Purali, an ancient name for Travancore

† In this we can recognize Uraiyur, the ancient capital of the Cholas

‡ This would seem to be the island of Mannar, formerly the headquarters of the Ceylon Pearl-fishery.

§ Probably written about the same time as the " Periplus "

The town of Sôsikourai has, I notice, not been identified by any writers with any now-existing place name, but I have little doubt it represents Tuticorin—the present principal town on the coast. The Tamil S is commonly corrupted into Ch, as Sippi into Chippi, and then Chochikourai would readily pass into Tôtikourai and Tûticourai from which Tuticorin, the present rendering of the name of this town, is readily derivable.*

Of Kolkhoi, identified by the Græco-Egyptian writers as the headquarters of the pearl fishery, the localization is rendered easy by reference to the Tamil poems of the period in question. In them is frequent mention of the great city of Korkai. Thus the *Maturaik-kanchi* describes it as the chief town in the country of the Parathavar and the seat of the pearl fishery, with a population consisting chiefly of pearl-divers and chank-cutters. The great epic *Chilappatikaram* (XXVII, 127)† further records that on account of the importance of the revenue derived from the pearl fishery, Korkai was a sub-capital of the Pandyan kingdom, and the usual residence of the heir-apparent. In its prime it boasted great magnificence, adorned with temples and palaces befitting its wealth and importance. It was situated at the mouth of the river Tambraparni, a river draining the present district of Tinnevelly and carrying down to the sea immense quantities of sand and mud.

The harbour of Korkai gradually silted up, and deltaic accumulation eventually cut off ready access to the sea. In consequence the old city gradually decayed and the population drifted to the new mouth of the river where a daughter town sprang into being at the expense of the parent. Exactly when this occurred I cannot make out‡ Mediæval travellers make no mention of Korkai or

---

* Tuticorin is the European equivalent or corruption of the Tamil name Tûttukkudi. The cerebral d often (usually) becomes r in the mouths of Europeans and the final n was added for the same euphonic reason that turned Kochchi into Cochin and Kumari into Comorin. Tûttukudi then becomes Tutikourai to the European ear and from this to the Sôsikourai of the Greeks is no difficult transition. A Greek mariner sailing along the coast of the Kolkhic Gulf could not possibly omit mention of the only sheltered port on the west side, apart from the river port of Korkai (Kolkhoi).

† சிலப்பதிகாரம், one of the five epics written by Tamils who were Buddhists or Jains by religion. It is known in English as "The Lay of the Lost Anklet." Its author was Ilanko adikal, brother of King Chenkkuduvan, an early Chera King who reigned probably about the end of the 1st or beginning of the 2nd century A.D.

‡ Korkai still exists as a name and a village. I visited it in 1912, and on the outskirts, within 20 yards of a stone partially embedded at the roadside, and bearing a sedant figure of a Jaina Tirthankar, I hit upon a rubbish heap from which I picked out a large quantity of chank bangle workshop refuse, similar to what is produced by Dacca shell-workers at the present day. Old Pandyan coins have been found at the same place

Kolkhoi : the headquarters of the Indian Pearl fishery still remained located at the mouth of the Tambraparni, * but its name was altered to Chayl, Cail, or Kayl, wherein we recognize the Kayal of to-day

Marco Polo in the thirteenth century speaks of Cail as a great and noble city ; Ludovico de Vathema mentions that he saw pearls fished for in the sea near the town of Chayl about A D 1500, while Barbosa, who travelled about the same time, says that the people of Chayl are jewellers who trade in pearls

To-day Kayal is a village some two miles inland and situated two and a half miles northward of Pinnakayal, a Parava town on an island in the present embouchure of the Tambraparni. The old name still clings, in the form Palayakayal, i.e Old Kayal, and mounds of rubbish littered with fragments of porcelain bespeaking the site of what must have been the great buildings of a noble city are within gunshot.

Kanakasabhai in his *Tamils of 1800 years ago* † states that Kayal and Korkai were separate cities, saying, "The site of this town (Korkai) which stood on the sea coast is now about five miles inland. After the sea had retired from Korkai, a new emporium arose on the coast. This was Kayal   . which in turn became in time too far from the sea and Kayal was also abandoned."

We have ample evidence that the abandonment of Kayal and the creation of the new ports and daughter towns of Kayalpattanam

---

* This name Tambraparni in its Romano Greek form of Taprobane was also the accepted cognomen of the island of Ceylon among the Romans of the empire. Variations in the manner of spelling are many—Tamrapurni, Tambraparni, Tambrapanni, Tamraparni and others

Much ingenuity has been displayed and wasted in seeking plausible derivations All those quoted in Tennent s ' Ceylon ' seem to be purely fanciful ; I do not think we need go beyond the terms *Tambiram*, copper, and *Varnam*, or *Parnam*, colour, words in common use among Tamils, in seeking for the meaning of the name

No feature strikes the stranger on arrival in Colombo more forcibly than the copper-red hue of the roads and soil ; "Copper-coloured Isle" is a most appropriate descriptive term to apply to Ceylon and equally so is the "Copper-coloured water" to the Tinne-velly river in question, when in flood it becomes turbid with the red mud it carries seawards

† Madras, 1904

2

and Pinnakayal (Pinnacoil), or Punnaikayal,* by the Moor and Parava inhabitants respectively of Kayal took place in early Portuguese times as will be noted further on.

When the Portuguese rounded Cape Comorin they found the pearl fisheries of the Gulf of Mannar in the hands of the caste of shore-dwelling people, fishermen and divers, already alluded to as Paravas whom tradition shows to have had control of this industry from time immemorial. Of the origin of these people we know extremely little. We know, however, that in the old days, from 600 B.C. and for 1,500 years or more thereafter, the country now comprehended in the districts of Madura and Tinnevelly, formed the great Tamil Kingdom of Pandya, and in the old Tamil work called the *Kalveddu* the position of the pearl-fishing caste to this monarchy is incidentally mentioned in the following extract:—

" Vidanarayanen Cheddi and the Paravu men who fished " pearls by paying tribute to Alliyarasani, daughter of Pandya, " King of Madura, who went on a voyage, experienced bad weather " in the sea, and were driven to the shores of Lanka, where they " founded Karainerkai (Karativo) and Kutiraimalai. Vidanarayanen " Cheddi had the treasures of his ship stored there by the Parawas, " and established pearl fisheries at Kadalihilapam (Chilavaturai) " and Kallachihilapam (Chilaw) and introduced the trees which " change iron into gold," etc., etc. (Herdman, " Report on the pearl oyster fisheries of the Gulf of Manaar," volume I, page 2.)

In the *Maturaik-kanchi* they are described as being most powerful in the country round Korkai. " well-fed on fish and flesh and armed with bows, their hordes terrified their enemies by their dashing valour.† " It is very probable that they were of Naga origin and of the same race as the inhabitants of Ceylon at the time of the Vijayan conquest of that island.

---

* Caldwell (*History of Tinnevelly*, page 72) derives Pinnakayal from *punnai*, the Indian laurel and *kayal*, a lagoon opening into the sea. This I consider most improbable, for this village is all but destitute of vegetation; even mangroves refuse to grow there. It is much more likely that the name has direct reference to the relationship of the place to its parent town, Kayal, the Cail of Marco Polo. The latter is known locally at the present time as Palaya Kayal (Old Kayal) and not simply as Kayal: this infers a daughter settlement somewhere, and hence I believe that Pinna Kayal means simply " New Kayal." De Sa e Menezes (vide footnote page 27) actually calls this place Puticale, which is the exact Tamil equivalent of " New Kayal " (Pudu kayal), and the Dutch sometimes used the form Pondecayl or Pondecail (see pages 170 and 171) having the same derivation—c.f. Pondicherry = Puducheri.

† Kanakasabhai, *loc cit*, page 44.

When the Pandyan kingdom was powerful the Paravas had grants of certain rights from the monarchy, paying tribute from the produce of the fisheries and receiving protection and immunity from taxation in return   The fishery in these early days appears to have been extremely prosperous—thus in A D. 1330 Friar Jordanus, who visited India at this time, tells us that as many as 8,000 boats were employed in the pearl fisheries of Tinnevelly and Ceylon.*

The efficient organization of the fishery was always a matter of great concern to the controlling power ; for example we learn from the following extract written by the Nawab of the Carnatic in 1771 to the Governor-General of Batavia, that

" In the time of the King of Madura, Terniel Nadu Raja † and
" the second king Minantje Ringeja Dalway, in the year 1470 it was
" decided that in January all things connected with the fishery
" should be arranged and that the same arrangement should hold
" good so long as the kingdom remained under the Carnatic.‡"

The conditions under which the Paravas lived and the far-reaching changes which in the middle of the 16th century were

---

* Thurston, E. " Pearl and Chank Fisheries ", Madras, 1894, page 9.

† This name evidently connotes that of Tirumalai Nayakkan, the greatest of the Madura Nayakkan dynasty. He reigned from 1623 to 1659   The date given for the transaction chronicled in this statement, 1470, is therefore obviously an error as no other Nayakkan ruler of similar name is known. The honorific *Nadu* given in the text, is a corruption of the Telugu *Nayudu*, an alternative of *Nayakka*   If therefore this regulation of the Pearl Fishery by the Madura ruler did take place, it was the work of the powerful and enlightened Tirumalai Nayakka (or Nayudu) aided apparently by his revenue minister, one of the Dalavay Mudali family ; the date must therefore have been between 1623 and 1659   The probabilities are all in favour of this ; the names are in agreement, while the coincidence of a powerful and ambitious ruler in Madura and the decadence of Portuguese power renders the interference of the former to recover the traditional authority of Madura over the Pearl Fishery extremely probable   It is exactly what we should expect

The person named as the second King was probably the Dewan or administrative adviser to the Nayakkan   Or he may have been only the farmer of the revenues, and as such directly concerned with obtaining a larger share of revenue from the Pearl Fishery. Dalavay was the hereditary honorific of a great family belonging to the Mudali or Mudaliyar division of the powerful Vellala caste ; members of this family farmed the revenues of Tinnevelly and Madura under the Pandyans and the Nayakkans ; even under the Nawab of the Carnatic they maintained this position ; they were men of vast import-ance   Their division of the Vellala caste occupied generally towards the rulers of the extreme south of India, both great and small, the same important role as administrative advisers that the Brahmans did elsewhere.

‡ Originally the term Carnatic meant the country inhabited by Kanarese-speaking people, the southern part of the great tableland between the Eastern and Western ghauts   Later, when the Nawab of the Carnatic, who succeeded to much of the power previously wielded by the great Hindu state of Vijayanagar, extended his territory to the Coromandel coast and the western shore of the Gulf of Mannar, this term " Carnatic" came to have a secondary and corrupt significance among English people, being limited to the lowlands between the Eastern ghauts and the sea.

beginning to be felt owing to the weakening of the paramount power of Vijayanagar are graphically set forth in a report, dated 19th December 1669, written by Van Reede and Laurens Pyl, respectively Commandant of the Coast of Malabar and Kanara and Senior Merchant and Chief of the sea-ports of Madura, in justification of their action in undertaking war with the Nayak or King of Madura This report addressed to Van Goens, the Governor of Ceylon and Dutch India, contains the following exposition of the condition of the Paravas prior to the arrival of the Portuguese, and of the manner in which the Portuguese obtained possession of the fisheries and subsequently carried them on :—

"Under the protection of those Rajas there lived a people
' which had come to these parts from other countries*—they are
' called Parruas—they lived a seafaring life, gaining their bread
" by fishing, and by diving for pearls ; they had purchased from
" the petty Rajas small streaks of the shore, along which they
" settled and built villages, and they divided themselves as their
" numbers progressively increased.

"In these purchased lands they lived under the rule of their
" own headmen, paying to the Rajas only an annual present, free
" from all other taxes which bore upon the natives so heavily, looked
" upon as strangers, exempt from tribute or subjection to the Rajas,
" having a chief of their own election, whose descendants are still
" called Kings of the Parruas,† and who drew a revenue from
" the whole people which in process of time has spread itself from
" Quilon to Bengal ‡ Their importance and power have not been
" reduced by this dispersion, for they are seen at every pearl fishery
" (on which occasions the Parruas assemble together), surpassing
" in distinction, dignity and outward honours, all other persons
" there, and still bearing their own appellation

"The pearl fishery was the principal resource and expedient
" from which the Parruas obtained a livelihood, but as from their

---

* From anthropometric and physical evidence this is probably correct The Paravas are distinctly brachycephalic whereas the Dravidians who constitute the higher castes in South India are notably long-headed and approximate closely in physical characteristics to the Mediterranean race. The Paravas are probably derived from ancient racial elements akin to the progenitors of the Polynesians—perhaps the Nagas of the ancient Tamil classics (See the author's papers on " The Outrigger Canoes of Indonesia " in the *Madras Fisheries Bulletin*, Vol 12, 1920, and " The Origins and Ethnol. Significance of Indian Boat Designs' in the *Mem Asiat Soc. of Bengal*, 1919).

† The Jathi Talaivan.

‡ Bengal This is, of course, not the Bengal of the Ganges delta, but the obscure Vangali, a village situated a few miles south of Mannar, on the Ceylon mainland.

" residence so near the sea, they had no manner of disposing of
" their pearls, they made an agreement with the Rajas that a
" market day should be proclaimed throughout their dominions,
" when merchants might securely come from all parts of India, and
" at which the divers and sutlers necessary to furnish provisions for
" the multitude might also meet, and as this assemblage would
" consist of two different races. namely, the Parruas and subjects
" of the Rajas, as well as strangers and travellers, two kinds of
" guards and tribunals were to be established to prevent all disputes
" and quarrels arising during this open market, every man being sub-
" ject to his o vn judge, and his case being decided by him, all pay-
" ments were then also divided among the headmen of the Parruas,
" who were the owners of that fishery, and who hence became rich
" and powerful; they had weapons and soldiers of their own, with
" which they were able to defend themselves against the violence of
" the Rajas or their subjects.

   " The Moors who had spread themselves over India, and
" principally along the coasts of Madura, were strengthened by the
" natives professing Mahomedanism and by Arabs, Saracens,
" and the privateers of the Sammoryn,* and they began also to
" take to pearl-diving as an occupation, but being led away by ill-
" feeling and hope of gain, they often attempted to outreach the
" Parruas, some of whom even they gained to their party and to
" their religion, by which means they obtained so much importance,
" that the Rajas joined themselves to the Moors, anticipating great
" advantages from the trade which they carried on and from their
" power at sea; and thus the Parruas were oppressed, although
" they frequently rose against their adversaries, but they always got
" the worst of it, until at last in a pearl fishery at Tutucoryn, having
" purposely raised a dispute they fell upon the Moors, and killed
" some thousands of them, burnt their vessels, and remained
' masters of the country, though much in fear that the Moors, joined
" by the pirates of Calicut, would rise against them in revenge.

   " The Portuguese arrived about this time with one ship at Tutu-
" coryn, the Parruas requested them for assistance, and obtained a
" promise of it on condition that they should become Christians;
" this they generally agreed to, and having sent Commissioners
" with some of the Portuguese to Goa, they were received under
" the protection of that nation, and their Commissioners returned

   * The Zamorin of Calicut, a powerful sea-chief of this period, but himself belonging
to the Hindu religion.

" with priests, and a naval force conveying troops, on which all the
" Parruas of the seven ports were baptized, accepted as subjects
" of the King of Portugal, and they dwindled thus from having their
" own chiefs and their own laws into subordination to priests and
" Portuguese, who however settled the rights and privileges of the
" Parruas so firmly, that the Rajas no longer dared interfere with
" them, or attempt to impede or abridge their prerogative, on the
" contrary they were compelled to admit of separate laws for the
" Parruas from those which bound their own subjects   The Portu-
" guese kept for themselves the command at sea, the pearl fisheries,
" the sovereignty over the Parruas, their villages and harbours,
" whilst the Naick of Madura, who was a subject of the King of the
" Carnatic,* made himself master at this time of the lands about
" Madura, and in a short time afterwards of all the lower countries
" from Cape Comoryn to Tanjore, expelling and rooting out all the
" princes and land proprietors, who were living and reigning there,
" but on obtaining the sovereignty of all these countries, he wished
" to subject the Parruas to his authority, in which attempt he was
' opposed by the Portuguese, who often, not being powerful enough
" effectually to resist, left the land with the priests and Parruas and
" went to the islands of Mannar and Jaffnapatam, from whence
" they sent coasting vessels along the Madura shores, and caused
" so much disquiet, that the revenue was ruined, trade circumscribed
" and almost annihilated, for which reasons the Naick himself
" was obliged to solicit the Portuguese to come back again

    "The Political Government of India, perceiving the great benefit
" of the pearl fishery, appointed in the name of the King of Portugal
" military chiefs and captains to superintend it, leaving the churches
" and their administration to the priests   Those captains obtained
" from the fisheries each time a profit of 6,000 rixdollars for the
" King, leaving the remainder of the income from them for the
" Parruas, but seeing they could not retain their superiority in that
" manner over the people, which was becoming rich, luxurious,
" drunken with prosperity, and with the help of the priests, who
" protected them, threatening the captains, which often occasioned
" great disorders, the latter determined to build a fort for the King
" at Tutucoryn, which was the chief place of all the villages, but the
" priests who feared by this to lose much of their consequence as
" well as of their revenue insisted that if such a measure was

---

* Vijayanagar is meant, at this particular period

" proceeded with, they would all be ruined, on which account they
" urged on the people to commit irregularities, and made the
" Parruas fear that the step was a preliminary one to the making
" all of them slaves ; and they therefore raised such hindrances to
" the work that it never could be completed.

"We have considered it worth while to prefix to our narrative
" this notice of old times, because it may throw some light on the
" present difficulties, and afford also a clear proof of the right which
" the Honourable Company at present claims over the Christian
" natives and all that relates to them.

" The Netherlands East India Company began about the year
" 1644, when it had obtained possession of some places in Ceylon,
" to carry on trade and commerce with the countries of Madura,
" and made a treaty for that purpose with the abovementioned
" Naick, stipulating,

" That the Honourable Company might trade in his territories
" with security and freedom, to which end a dwelling or lodge at
" Cailpatnam was allowed them, as may be seen by the treaty or
" contract in possession of the Company—and on this word and
' faith of the Naick their trade began, and their goods, merchandize
" and servants were confidently left in protection of their ally—but
" where there is no firm ground of integrity, treachery and faith-
" lessness find easy entry.

"This the Company soon experienced, for it was not long ere
" this evil-minded and wicked people, deceived by appearances,
" and induced by hope of rapine and profit, forgot their faith and
" promises, suffering themselves to be seduced by a sum of money
" to demolish the Company's lodge, seize its goods, and murder its
" servants ; in which last attempt however they failed by the
" unexpected appearance of a ship in which the men took refuge
" and thus wonderfully escaped. This dastard villany, detestable
" in any prince or chieftain, the Portuguese had contrived, and
" effected by means of the Parruas and the Naick's servants who
" thought the neighbourhood of the Company injurious to their
" interests And although in the year 1649, a signal vengeance
" fell as well upon Tritchenadoor as Tutucoryn, yet the people, and
" their master, the Naick and his Government remained equally
" base, taking every opportunity to exercise oppression Even the
" Portuguese whom they had assisted to do harm to the Company
" began soon to perceive that the renters and chiefs of the lowlands
" wronged the Parruas on all sides and diminished the right of the

" Portuguese, but at that time they had no means of preventing it,
" as the Dutch Company was increasing in strength and was
" taking possession of their towns, forts and ships, and became
" daily more powerful, which caused them to bear much from the
" Naick with forbearance.

" Matters stood thus in the lands of the Naick of Madura* till 1658,
" when the town of Tutucoryn was taken by force of arms from the
" Portuguese and Parruas, by which success the Company succeeded
" to their rights over the coast, as well as to their authority over the
" sea-ports, the Christians, the pearl fisheries, and all thereunto
" appertaining ; in fact to all that the Parruas first had, and the
" priests and Portuguese afterwards possessed."

(b) THE PEARL BANKS UNDER PORTUGUESE CONTROL,
1524—1658.

Of the prosperity and conduct of the fisheries under the Portuguese we know nothing with exactitude—even the dates of the
important fisheries are lost through the disappearance of the official
records.

---

* Dravida, the country of the Tamils, was divided in the earliest days of which we
have record and prior to the Christian era, between three dynasties, the Pandyans, the
Cheras, and the Cholas. The Pandyan kingdom, deriving its name from that of the
founder of the first dynasty, comprised under normal conditions little more than the present
districts of Madura and Tinnevelly, the city of Madura being the capital during the greater
part of the continuance of the kingdom, which suffered the usual vicissitudes of Indian
states, sometimes preponderating and more frequently in later times, tributary to a neighbouring state, but always maintaining in Madura some semblance of sovereign authority.

After the fall of the powerful Hoysala Ballala Rajas, at the beginning of the
fourteenth century (A.D 1310), a great Hindu state, that of Vijayanagar, took shape in
the centre of the Deccan. The sovereign of this state became about the middle of the
fourteenth century the overlord of the states of Southern India including the Pandyan
country, and the Princes of Madura remained tributary till about the time of the arrival
of the Portuguese in the Gulf of Mannar

The reigning dynasty at that time was that of the Nayakkans, and while   the
Portuguese were busy making settlements on the coast, the Nayakkan was making himself
master of all the lower countries from Cape Comorin to Tanjore, " expelling and rooting
out all the princes and land proprietors " who were living and reigning there. The
Nayakkans did not consolidate their power in the extreme south till after St F. Xavier's
arrival in India in 1542

The battle of Talikota in 1565, in which the King of Vijayanagar fell before a
great combination of Muhammadan states, gave the Nayakkan complete independence,
which his family retained till 1736 when the last of this house fell before the power of
the Nawab of the Carnatic, the ally of the British

It is noteworthy that the Nayakkan rulers of Madura never actually assumed a regal
title, contenting themselves with their original one (Nayakka) meaning Lieutenant or
Viceroy, even after the suzerain power of Vijayanagar was broken by the Muhammadans,
and when, as in 1639 there was actual war between the reduced Raya of Vijayanagar
(then of Chandragiri) and the Nayakkan of Madura,

Fortunately many of the Portuguese soldiers of fortune have left memoirs of their lives in the east, and several furnish interesting accounts of the conduct and management of the fisheries at this time. The two most important are those left by Gaspar Correa and by Juan Ribeyro; the former dealing with the condition of the fishery at the beginning of the Portuguese connection, the latter with an account of it in the years when his nation was being dislodged foot by foot and fort by fort from Ceylon and India by the Dutch. Both are worthy of being better known and therefore it will be better to give the extracts in full

Gaspar Correa's account tells [*] us that in the year 1523 the King of Portugal commissioned Manuel de Frias to make inquiries in India regarding the tomb of the apostle Thomas, and it proceeds thus :—"And he commanded that with him should go João Froles, "who had taken part in the affairs in Ceylon when Lopo Soares "went there, and who had been appointed by the King captain and "factor of the pearl fishery which is carried on by the natives of "the country between Ceylon and the Cape of Comoryn ,[†] in former "times the Moors of that coast had possession of this pearl fishery, "for which they paid a large rent to the lords of the land, where- "fore the Governors had a good right thereto, since they were "rulers of the sea Therefore now João Froles having come thus

---

[*] " Lendas da India '

[†] From Correa it would appear that it was *circa* 1525 that the Portuguese seized the pearl fishery on the *Indian* coast (this is to be inferred as his narrative tells of the Moor pirates *returning to Ceylon* after the sea-fight described on the next page) It does not appear that they settled on the Tinnevelly coast at this period ; they probably did little except raiding the fishery when they heard that one was in progress—holding it to ransom and squeezing what profit they could out of the merchants and divers This will explain Caldwell's statement that the settlement (seizure) of this coast by the Portuguese took place in 1532. He was apparently not aware of the raiding of the fishery in 1525 and 1528 According to his account (*History of Tinnevelly*, page 68), in 1532 a deputation of Paravas went to Cochin to pray for aid against the Muhammadans The deputation is said to have comprised 70 persons The appeal was granted and an expedition was fitted out The idea of seeking this aid and the offer of the caste, which was made at the same time, to embrace the Roman Catholic religion, is credited to the advice of a native convert, João de Cruz. Father Michael Vaz, the Vicar-General at Cochin, accompanied the expedition with some priests and on his arrival on the coast, after the overthrow of the Muhammadans, 20,000 Paravas, occupying 30 villages, are said to have been baptized St F Xavier stated some years later that the Chiefs of the Saracens were slain and their power utterly broken. Father Michael Vaz was described by Xavier as " the true father of the Comorin Christians " Xavier arrived in 1542 and laboured on this coast two years Up to the last he seems never to have been able to speak Tamil, though he learned off by heart certain prayers and formulæ

"commissioned that he might take possession of and receive for
"the King this fishery. so that the Governor might not suffer loss
"by not being able to receive it for himself, he (the Governor of
"Portuguese India) did not give João Froles the fleet and men that
"the King had commanded, and in order to get from it whatever
"he could gain he ordered Manuel de Frias to go to the fishery,
"and to rent it for whatever the lords of the land would give, and
" this in order that he might find out what it would yield, and
"having accomplished this that he should proceed to the coast of
"Choromandel as captain and factor."

A little further on we read.—"Manuel de Frias, captain and
"factor of Choromandel, in accordance with the orders of the Gover-
"nor, which he carried with him, placed João Froles over the fishery
"which he rented out to the *digares,* * of the country for one thousand
"five hundred *cruzados* per annum, and left there as factor João
"Froles, with his clerk, in a large boat well armed, and as
"the factor was not able to steal any of the money from the
"rent of the fishery, he took other measures, obtaining from the
"fishers themselves the pearls, whereby he committed many
"robberies, as is done nowadays; for the ills of India are not
"improving, but are increasing continually, as I shall recount
"further in speaking of the end that this João Froles came to at
"this fishery, in which he paid for a portion of the evils that he
"had committed"

An interval of three years then elapses before we again hear
of the pearl fishery, Lopo Vaz de Sampayo being then Governor
in India. We read next that in January 1528.—

" Manuel da Gama was appointed by the King as captain of
"the coast of Coromandel, and João Froles as captain and factor
"of the pearl fishery. This act of friendship towards Manuel da
"Gama was managed by Hector Da Silveira before he departed,
"and the Governor gave him a ship and four foists well fitted
"and armed, as he had had tidings that paraos of Calicut were
"going along the coast of Paleacate (Pulicat)† committing great

---

* Evidently *ddigars* are meant. At the present day the principal headman of the
pearl fishery district in Ceylon (Musali) is officially known by this title of Adigar. It is
this officer who is charged with the details incident to the erection of the fishery camp
prior to each fishery

† The Portuguese appear to have established themselves at Pulicat, 27 miles north of
Madras, between 1522 and 1525.

" robberies, and had seized a ship that had come from Malacca
" very richly laden, with eight Portuguese whom they put to death.
" To this Manuel da Gama retaliated so well, that he cleared the
" coast of the robbers, and managed to get back on land all the
" goods from the ship which the robbers had sold, and many male
" and female slaves of the Portuguese whom they had killed on
" board the ship ; which robbers went over to Ceylon with much
" booty, and joined the others who had gone from Calicut, and
" went about robbing as much as they liked by sea and land.
" The Governor sent João Froles as captain and factor of the
" fishery, in a caravel and a large boat and three foists, with
" which he went about collecting the rent of the fishery, as I have
" already said. This being known to the robbers, who went about
" strongly armed with artillery and men, twenty of them (paraos)
" came in a body to attack João Froles as Manuel da Gama had
" gone to the other coast and could not help him and they
" came upon João Froles who was in the caravel, with the large boat,
" the foists having gone to another place ; and as they were moored
" and the wind was calm, twelve of the paraos made for the caravel,
" dividing into six on each side, and the other eight likewise
" divided to attack the large boat. João Froles, seeing the paraos
" preparing for the attack, made ready as well as he could with
" twenty Portuguese men that he had, and threw a rope to the
" large boat, so that the two lay stern to stern. Six Portuguese
" men went into the large boat . the caravel had a *camello* and two
" falcons and six *bercos*,* and the large boat, two falcons and six
" *bercos*, but there were only a few men as several had gone to the
" foists that João Froles had sent to the coast of Ceylon as prizes.
" Our men having thus got ready, the *paraos* divided into two
" attacking parties, ten approaching from each side avoiding the
" shots from the camello, and shifting as they pleased, all the
" while discharging from *roqueras** iron balls of the size of quinces,
" and firing as they liked they gave the caravel and large boat so
" many shots, that they cut their shrouds and caused them to fall
" with the yards, at which they set up loud shouts. Neither João
" Froles nor the master of the caravel had thought of putting belts
" under the yards, which if they had done the yards would not
" have fallen At this the Moors considered themselves victors,

----

* Different kinds of cannon

" as the Portuguese were already killed or wounded, for only the fal-
" cons and *bercos* were now of any use to our men, and they did not
" fire them so often as did the Moors, and our men were continually
" becoming less able to fire; wherefore the Moors knowing the
" weakness of our men came in a body with their arms, and their
" shouts and war charges, and boarded the vessel, and killed as
" many as they found alive, without sparing any one, and carried
" off all that they found, and took the falcons and *bercos* and
" ammunition, and set fire to the vessels so that they went down,
" and then returned to Ceylon. Our foists, hearing the news of the
" burning of those vessels, fled to where Manuel da Gama was
" staying."

The description of the conduct of the fishery under the Portuguese by Juan Ribeyro in his " History of Ceylon " dated 1685, is
the only detailed account handed down to us.*

It runs as follows :—

" Having now related all that we know of the natural riches
" of the land of Ceylon, we shall describe those which its sea
" produces. The pearls which are procured from the coasts of
" the island, and more especially from Aripo, are of the highest
" value As few persons know how that fishery is conducted, we
" shall here relate what we know of it.

" At the beginning of March there assemble on that coast 4,000
" or 5,000 boats got together and paid by Moorish or Heathen
" merchants and by some Christians.† These merchants have
" many partnerships among themselves, and they first make up a
" fund to arm four, five or six boats, more or less, according as
" the entire adventure is greater or smaller. Each of these boats
" has generally from ten to twelve sailors, one master and eight or
" nine divers. All the boats go out together, and seek where
" the fishery is likely to be most profitable and they anchor at the
" spots where the sea is only five, six or at most seven fathoms
" deep Then they send off three boats to a league distant round
" about, each in a different direction ; each of these boats brings

---

* From the English translation from the French version of the Abbe Le Grand,
Ceylon, 1847.

† " An escort of armed men always accompanies the pearl divers, on account of the
Malabars, who come from the coast of that name or from the Maldives, and who live by
piracy so that no boat, canoe or prahu is safe in those seas. The fishers or divers cease
their work at noon, on account of the swell caused by the wind, and which annoys the
divers, who can only descend in calm weather." (Note by the French translator.)

" back a thousand oysters These are opened in presence of the
" merchants and the pearls found in them are examined by the
" whole party and their value estimated, as the pearls are much
" finer in some years than in others, and accordingly as the
" merchants find the pearls to be large, clear, round and of good
" water, they bargain with the King for the fishery of that year
" When the bargain is made the King usually gives them four ves-
" sels of war to defend them from the Malabar and other pirates.
" Then each merchant goes to the sea-side and constructs a sort of
" enclosure with stakes and thorns, only leaving a narrow passage
" for the boats to enter and go out again, which come there to
" discharge the oysters they have fished up

　　" On the 11th of March, at four in the morning, the officer in
" command of the four vessels of war fires a gun as signal, and
" immediately all the boats put off to sea, steering for the place
" which they have selected to fish at and casting anchor there
" Each of these boats has on board stones of the weight of 60 lb
" each, fastened with strong ropes, of which one end is attached to
" the boat The diver places his foot on one of the stones, and
" passes another rope round his body, to which is tied a basket or a
" small woven bag like a net; this second rope is held by two of
" the sailors, and the diver thus secured descends into the sea; he
" remains there whilst two *credos* can be said, and fills his little
" bag or basket with oysters which he sometimes finds in heaps on
" the rocks, as soon as his basket is full, he makes a sign by
" pulling the rope held by the sailors in the boat, and one end of
" which is round his waist, and they draw him quickly out of the
" water; but if in the time he is below, he can contrive to open an
" oyster and finds a pearl in it, it is considered his own *; as
" soon as his head is above water another diver goes down, and
" thus they descend by turns This fishery lasts till four in the
" afternoon, when the officer in command fires another gun as a
" signal to cease the fishery for the day Then all the boats go to
" their several enclosures, and the noise and confusion that ensue
" in the two hours that are allowed to discharge and pile up the
" oysters cannot be described

---

　　* This is an error in translation and in fact In translation from the original
Portuguese it reads, " the diver as soon as he rises to the surface is at liberty, until
he who is at the bottom of the sea ascends, to open with a knife as many oysters
as he can and whatever he finds therein is his "

"Besides the people belonging to the boats the children of the
" neighbourhood never fail to assemble at the sea-side, offering
" their services, rather however to steal the oysters than to assist
" the sailors or merchants. As soon as the boats are unloaded they
" put to sea again, and go about half a league higher up by the sea-
" side, where the merchants assemble and hold a splendid fair,
" there are magnificent tents and all sorts of merchandise of the
" most valuable kind are to be had there, as vendors come from all
" parts of the world Heathens, Jews, Christians and Moors all have
" some speculation for profit; some sell by wholesale, others by
" retail ; the sailors and children bring the pearls which they have
" stolen, and people of every kind have bargains to offer Persons
" having but a small capital buy small ventures, which they
" immediately sell to larger merchants with a middling profit, not
" only pearls are bought and sold, but jewellery of every kind,
" bargold, dollars, fine Turkey carpets, and beautiful stuffs from
" India

"The fishery lasts from the 11th of March to the 20th of
' April,* but the fair itself continues for fifty days, because for the
" last nine days the enclosures are cleansed, as so many flies are
" bred by the corrupt matter that the adjacent places and the whole
" country might be annoyed by them, if care were not taken to sweep
" into the sea the impurities collected during the fishery.

"On the last day of April, the merchants of the several
" partnerships assemble together and share the pearls belonging to
" their respective boats. They separate them into nine classes, and
" set on each class a price according as the demand has been greater
" or less for pearls during the year, when these prices have been set
" on them, they make the allotments and shares Then the ill-
" formed pearls are sold at a sufficiently moderate price; the small
' seed-pearls are left on the sea-side and the country people come
" in the spring and sift the sand for them and sell them for a trifle

" Hence the pearls and seed are sent to all parts of the world
" This is all I know of this fishery. But I must not forget to add
" that pieces of *amber*† of a considerable size are also found on this
" coast Great branches of *coral* also drift ashore when the sea is
" high ; the black kind is better and more esteemed than the red."

---

* Old style ?

† Ambergris is meint. No amber is found in Ceylon or India

In no Portuguese work have I found any indication of the frequency of the recurrence of fisheries under the Portuguese or of the approximate value and locality of any, a lack of knowledge greatly to be regretted as it becomes impossible to say with certainty whether or not there has been deterioration, progressive or intermittent, in the oyster-producing qualities of the beds The only hint I have come across is a chance remark in Ribeyro's " History ", to the effect that the inhabitants of Mannar had in his time (circa 1658) become impoverished by the decadence of the pearl fishery on the Ceylon coast and its transference to the Tuticorin side, his words being " at present the oysters have migrated and are to be found on the coast of Tuticorin."*

Even prior to the Portuguese we find the uncertainty of the pearl fisheries a matter of notoriety for Albyrouni who served under Mahmoud of Ghazni and wrote in the eleventh century, says that the pearl fishery which formerly existed in the Gulf of Serendib, had become exhausted in his time simultaneously with the appearance of a fishery at Sofala, in the country of the Zends, where pearls were unknown before, and remarks that hence arose the conjecture that the pearl oyster of Serendib had migrated to Sofala†, i e., to the Persian Gulf.

Few other facts of importance are to be gleaned from Portuguese writers. We see however that in fulfilment of the treaty made with the new-comers the Paravas become zealous Roman Catholics Thus they won the confidence of their masters and under the protection of the priesthood enjoyed a comparative tranquillity and immunity from extortionate tyranny seldom met with by Indians living within the Portuguese influence.

St. Francis Xavier did great work among the Paravas and it was on the fishery coast at or about Pinnakayal that he commenced his missionary labours in 1542, thereafter visiting Tuticorin

---

* See also Baldaeus' *Description of Malabar and Coromandel*, English edition, London, 1703, where in volume 3, page 792, referring to the condition of affairs in 1658, he states that "this island (Mannar) was formerly celebrated for the pearl fishery as well as the city of Tutecoryn ; but no pearls having been taken there for these ten years last past, the inhabitants are reduced to great poverty ; whereas the sumptuous edifices, churches and monasteries, with their ornaments, are sufficient demonstrations of its former grandeur "

† Reinaud's " Fragmens Arabes," page 125, quoted by Tennent in *The Natural History of Ceylon*, page 375

and sending priests to Mannar at the earnest solicitation of the inhabitants *

That the fisheries were then flourishing is betokened by the fine churches and great monasteries that rose at the three centres named from the offerings and profits of the divers and merchants during the second half of the sixteenth century

The Portuguese appear to have kept well in hand the petty Chiefs whose territories abutted on the fishery coast and to have been able to afford efficient protection to the Paravas They were fortunate in arriving in India at a time when the native states were in the crucible of change, when internecine warfare left the chiefs neither time nor power to cope effectually with a more highly organized foe from oversea. Old states were in the melting pot of invasion and insurrection and especially was this true of Southern India, where political paralysis began to affect Vijayanagar— beginning, as is usual, in those provinces furthest from the centre of the state.

The latest dynasty—the Nayakkan—occupying the tributary throne of Madura was beginning to assert independence of the central Government from which it became virtually free when the battle of Talikota in 1565 completed the destruction of the Suzerain Hindu State of Vijayanagar.

The Paravas as already mentioned, although the original holders of the fishery rights, had begun prior to the arrival of the Portuguese to feel the competition of the restless Muhammadan settlers on the coast, who, coming as many must have done, from the coasts of the Persian Gulf knew already all there was to know of pearl fishing The descendants of these Arabs and their proselytes, known as Moros to the Portuguese, are the Moormen or Lebbais of to-day †

Hating the Moormen with all the fanatical intolerance which was their curse and the chief cause of their eventual ruin, the Portuguese as we have seen took the part of the Paravas. Accordingly the Nayak of Madura, Hindu though he was, habitually lent his influence to the Moors in the hope of eventually being in a

---

* According to St Francis Xavier drunkenness among the Paravas was gross in his time, he was furious with the headmen or *pattangattis* for them free indulgence in arrack He threatened to have them sent in chains to Cochin if they did not reform.

† Moormen is the appellation used in Ceylon, whereas Lebbais is commonly used on the Indian Coast for the same people.

position to drive out the Portuguese and so obtain control of the coveted pearl fishery. None of the Nayaks was ever strong enough to do this and an armed neutrality usually existed, the Portuguese even granting certain privileges—practically tributes—to the Nayak's Government in return for facilities given to pearl merchants to travel without exactions to the scene of the fishery.

The chief item in the concession made to the Nayak was the grant of a number of free boats in each fishery. A grant engraved upon copper purporting to be made by Tirumalai Nayakkan in favour of the Mudaliyar Pillai Marakkāyar, the head of the Moorish community,* on founding or re-settling the town of Kayalpattanam furnishes an interesting light on the details of this arrangement.

It appears that a new port at the mouth of the Tambraparni, free from the domination of the Portuguese, had become a necessity to the Nayak through the silting up of the harbour of the city of Kayal by sand brought down by the river.

In recognition of the headman's enterprise in settling a large number of his people at Kayalpattanam and thus conserving to the Nayak a sea-port able to rival the Pinnakayal and the Tuticorin of the Portuguese, several gifts were made to him the chief being the grant of ten "free" divers' stones at the fishery. In return he was with "seven large boats, with 96½ stones, at 13¾ stones to each boat, to fish the pearl banks for the use and benefit of the said Government" (of Madura) It is expressly said, "he is to

---

* Termed "Choliars" in this grant. In the tenth century the Chola dynasty overthrew the neighbouring sister kingdoms of the Chera and Pandya, and reigned paramount from the vicinity of Madras to Cape Comorin

It was doubtless subsequent to this period that the Tamil Muhammadans of South India became known as the *Choliya Muhammadans* or more commonly *Choliyar* or people of the Tamil country called *Chola desam*. To this day the Hindustani Muhammadan speaks of his southern co-religionist as Choliya ; for, save as to religion the vast majority of the Choliyar are Tamils in point of language, general appearance and social customs—*vide* Ramanathan "The Ethnology of the Moors of Ceylon" in *Journal R.A S (Ceylon Branch)*, Volume XIII, page 245.

In the seaports, particularly Kilakarai and Kayalpattanam, these vernacular-speaking Muhammadans contain a marked strain of Arab blood in the higher class which usually arrogates to itself the honorific suffix of Marakkayar. These men usually know something of Arabic, or at least can write Tamil in Arabic characters. Occasionally a purely Arab type of features and physique is met with among them, as in the case of the late Ahmed Jalalludin Marakkayar, chank-fishery renter and shark-charmer, of Kilakarai. The lower classes have the features and physique of the local coast people of the place they inhabit On the west coast, the Mappillas, of similar two fold origin, replace the Lebbais of the south-east coast

4

reside near the Government House of the Portuguese at the sea-port of Mannar and near Mári Amman's chapel at Tuticorin.* *He shall have* the superintendency of the pearl fishery and shall receive 60 chacrums per month and shall be favoured with ten stones to dive for him at the said two places" (Mannar and Tuticorin).

The 96½ stones above mentioned represent the allowance conceded by the Portuguese to the Nayak *in return for the privileges before named.* Later we shall see that the question of the consideration given in return for this privilege became the source of continual disputes between the Dutch and the Nawab of the Carnatic, the latter succeeding by conquest to the rights of the Madura Nayaks in the early part of the eighteenth century.

Other free stones† were at intervals during the sixteenth century granted out of these 96½ privilege stones by the Nayak to various temples from religious motives, as in 1542 and 1546

Besides the Nayak of Madura, the Portuguese allowed to his tributary, the Setupati of Rāmnād, a further number of free divers (60 stones) in each fishery in return for the help he gave in contributing to the success of the fishery and in guarding and providing pilots for the passage of the narrow strait called Pāmban pass, separating the mainland from the Island of Rameswaram.

This petty sovereign, who is the hereditary guardian of the temple of Rameswaram and is the head of the Maravar caste, was commonly known as the Theuver or Tuever in the days of the Portuguese and the Dutch. While nominally under the Madura monarch, the Setupati was virtually independent and leaned more to the foreigners, for his lands being coastal and insular, danger was greater from the sea than from the land. His territory included the coast as far south as Kilakarai, the great Moor diving centre at the present day, and for that reason his assistance had to be courted and purchased by the European lords of the pearl fisheries.

---

* The Portuguese made Tuticorin their chief settlement on the Pescaria coast about 1580. Prior to that Pinnakayal was their headquarters The fire Goa church at Tuticorin, dedicated to N Senhora das Nieves (in Tamil *Pani maya-mata*, dew replacing snow) was founded in 1583 (Caldwell, *History of Tinnevelly*, pp 75 to 78)

† By "stones" divers are to be understood, a diving stone being the indispensable item in a diver's equipment. Each stone is however, usually shared by two divers, so it is probable that the 96½ stones here referred to, represented an allowance of 193 divers

Like him of Madura, the Rāmnād lord* granted some of his privilege divers to the great Hindu temples of his district, giving seven stones to Rameswaram pagoda in 1609 and three more in 1714

Besides the 60 free stones, the Setupati had the right by custom under the Portuguese to one day's fishing from all his subjects, as had the Nayak from his

Taken generally the fisheries under the Portuguese appear to have been of great collective profit during the first half of the period of their rule, a period coincident with the height of Portuguese energy and power, when they had no European rival and when they were free to concentrate their forces entirely against the native races After breaking the power of the Arabs, the Portuguese enjoyed all the advantages conferred on the nation possessing the mastery of the sea, a consideration of supreme importance in connexion with such an essentially maritime industry as the pearl fishery

Encroachments and claims on the part of the Nayak of Madura were then as common and as troublesome as those experienced in the eighteenth century by the Dutch in their relations with the Nawab of the Carnatic. The methods adopted by the Portuguese to cope with these infringements of treaty rights and to afford

---

* The Chief of Rāmnād has many titles indicating an ancient and illustrious past. To-day his lands form a zamindari , prior to the permanent settlement made in 1803, Rāmnād was a semi feudal state (*palaiyam*) dependent upon Madura, but occasionally, when the central power was weak, throwing off the yoke and assuming virtual and even actual independence as evidenced by the existence of coins struck by certain of its rulers The present Zamindar of Rāmnād is styled Raja, a distinction conferred by Government in recognition of his public spirit ; in former days Setupati (or Sethupathi) and Tevar were titles more particularly distinctive and peculiar to these chieftains

The Raja of Rāmnād is the hereditary head of the warrior Maravar caste, of which the honorific and generic caste name is Tevar (or Thevar) ; from this was derived the title Tevar, by which the Dutch referred to the Sovereign of Rāmnād in all their documents, usually under the form Theuver or Teuver.

The Rajas of Rāmnād have always been identified with the great Hindu shrine of Rameswaram , they are its hereditary guardians Their ancient title of Setupati is connected directly with this honourable distinction, for in its meaning of ' Lord of the Causeway '' it connotes the guardianship of the sea between Ceylon and India, and of Adam's Bridge, the line of islets and sandbanks connecting the islands of Rameswaram and Mannar The Setupatis, in virtue of this guardianship, held possession also of the narrow channel, known as Pamban Pass, between Rameswaram Island and the mainland and levied dues on vessels passing through

In the settlement of 1803, while the sovereign rights over any future pearl fishery off the Rāmnād coast were retained by the paramount power, those over the chank fishery—an industry of considerable annual value—were included among the] privileges accorded to the zamindar.

protection to their subjects and allies, the Paravas, appear to have been most radical and effective, consisting in the removal of all Christian natives from the Madura coast to Mannar and to the string of islands skirting the coast from Tuticorin to Pāmban with a concurrent blockade of the Nayak's seaboard. Nor was the blockade a peaceful one as we learn from Van Reede and Laurens Pyl's Memoir of 1669 quoted above. To use their words "the Portuguese with their boats pillaged the entire sea-coast, which they disquieted so effectually that the renters and overseers (of the Nayak) on account of the great loss they suffered in their revenues were obliged to request the Nayak to call the Portuguese back again. "

During this period of disturbance the Paravas held pearl fisheries from the small islands along the Madura coast and "assisted to the best of their power the Portuguese vessels. " *

I refer to this period of the temporary settlement of the Paravas in the Madura Islands, the unmistakable evidence of a fishery camp to be seen to-day among the sand-dunes of Nallatanni tivu, an island lying off the coast between Kilakarai and Tuticorin. If we fix the date of this fishery between 1560 to 1570 we cannot be far out, for it was in 1560 that the Viceroy Don Constantine de Braganza erected the fort of Mannar and transferred thereto the inhabitants of the Parava town of Pinnakayal, the scene of Francis Xavier's labours twenty years previously, and one of their chief and most prosperous settlements.†

By this change the island of Mannar become rich and prosperous as long as the fisheries continued to give handsome returns. De Sa e Menezes (loc. cit.) writing in 1622 states that for many years the fisheries had become extinct "because of the great poverty into "which the Paravas had fallen, for they made no profit for want of accommodation and of boats "—a result likely to arise from the

---

* Van Reede and Pyl, loc. cit.

† De Sa e Menezes describes Pinnakayal (Puticale as he spells the name) at the time of this transference as " a place on the Fishery Coast, inhabited by Parawas, who, tired of the continual attacks of the Bodaguas, their neighbours, lived more the life of fronteros than of fishermen, which trade they plied for subsistence, but were continually robbed and cut off by their neighbours "—The Rebellion of Ceylon, translated from the Spanish by Lieutenant-Colonel H H. St George). These Bodaguas (properly Vadugas) were the Nayakkan's Telugu taxgatherers, so called in the Tamil country because, being Telugus, they came from the north. They belonged to the ame caste as the Nayakkan himself. A Jesuit writer of that time, quoted by Caldwell (loc. cit. p 69), described them as "the collectors of the royal taxes, a race of over-searing and insolent men."

exactions of Church and of State, from the natural improvidence of
the race and from the rapid decay of the Portuguese sea-power
consequent upon the successful inroads made upon their monopoly
of sea-borne commerce between India and Europe. The Portuguese, struggling for very existence and in continual straits for the
money requisite to carry on an exhausting contest, increased their
exactions from the natives and at the same time were unable to
give them adequate protection, especially at sea. We may infer
with every probability of this being true, that from the time the
Dutch appeared in force in Indian seas, a time coinciding with a
period of great official corruption and internal unrest among the
Portuguese, the management of the pearl banks became inefficient
and badly conducted

(c) THE PEARL BANKS UNDER THE DUTCH, 1658–1796.

Tuticorin and the sovereignty of the pearl banks and of the
Paravas passed to the Dutch in 1658. In 1663 the first fishery
under the new rule was held, resulting in a profit of 18,000 florins.*

At this fishery the Nayak of Madura and the Setupati of
Rāmnād and the head Moor of Kayalpattanam had their accustomed number of boats free as under the Portuguese.

Just prior to this fishery Cornelis Valkenburg had written " The
"fishery of Mannar (Gulf of Mannar) is in great repute with the
"Portuguese and everybody else, but if it be really of much
"importance has not yet been experienced and therefore I can
" give no information on the subject. "†

The second Dutch fishery took place six years later, in 1669,
with what profit I do not know. Then a long interval of 22 years
occurred bringing us to the third fishery, 1691, at which there were
385½ stones admitted free, viz. : –

96½ for the Nayak of Madura

59 for the Setupati of Rāmnād and the remainder for the
headmen of the divers, divided on the lines detailed in the statement of these arrangements at the Ceylon fishery of 1694 appended.

Six years later, 1697, we find Croon, the Commandant of Jaffna,
writing that "the pearl fishery is an extraordinary source of
revenue, on which no certain reliance can be placed, as it depends

---

* See Appendix A, page 171.
† In instructions left for the guidance of his successors, the Residents of the
Seven Harbours on the Madura Coast, *Ceylon Lit. Reg.*, Vol. III, page 160.

on various contingencies which may ruin the banks or spoil the
oysters. If no particular accident happen, it may take place for
years successfully   .   .   but if the oysters happen to be
washed off the banks or to be disturbed by storms, the banks may
be totally ruined in a very short time   The examination
. . . is superintended by Commissioners specially appointed
and is conducted in   dhonies by Pattangattyns and
other native headmen, who understand the business " *

The next glimpse we get of a fishery off the Tuticorin coast is
in the graphic description by Father Martin, a Jesuit missionary,
of a disastrous three days' fishery held in 1700. The description
in spite of errors in detail is so vivid and instructive that it may
well be reproduced here for comparison with that left on record by
Ribeyro of the methods pursued in Portuguese days and also with
those employed at the present time :—

"In the early part of the year the Dutch sent out ten or twelve
" vessels in different directions to test the localities in which it
" appeared desirable that the fishery of the year should be carried
" on; and from each vessel a few divers were let down who
" brought up each a few thousand oysters, which were heaped
" upon the shore in separate heaps of a thousand each, opened and
" examined   If the pearls found in each heap were found by the
" appraisers to be worth an *ecu* or more, the beds from which the
" oysters were taken were held to be capable of yielding a rich
" harvest; if they were worth no more than thirty sous, the beds
" were considered unlikely to yield a profit over and above the
" expense of working them.   As soon as the testing was completed,
" it was publicly announced either that there would, or that there
" would not, be a fishery that year.   In the former case enormous
" crowds of people assembled on the coast on the day appointed
" for the commencement of the fishery; traders came there with
" wares of all kinds; the roadstead was crowded with shipping,
" drums were beaten, and muskets fired; and everywhere the great-
" est excitement prevailed, until the Dutch Commissioners arrived
" from Colombo with great pomp, and ordered the proceedings to
" be opened with a salute of cannon   Immediately afterwards the
" fishing vessels all weighed anchor and stood out to sea, preceded
" by two large Dutch sloops, which in due time drew off to the
" right and left and marked the limits of the fishery, and when

---

* Lee's translation of Ribeyro's " History of Ceylon," page 247

" each vessel reached its place, half of its complement of divers
" plunged into the sea, each with a heavy stone tied (*sic*) to his feet
" to make him sink rapidly,* and furnished with a sack into which
" he put his oysters, and having a rope tied round his body, the
" end of which was passed round a pulley and held by some of the
" boatmen. Thus equipped, the diver plunged in, and on reaching
" the bottom, filled his sack with oysters until his breath failed,
" when he pulled a string with which he was provided, and, the
" signal being perceived by the boatmen above, he was forthwith
" hauled up by the rope, together with his sack of oysters. No
" artificial appliances of any kind were used to enable the men to
" stay under water for long periods; they were accustomed to the
" work almost from infancy, and consequently did it easily and
" well. Some were more skilful and lasting than others, and it
" was usual to pay them in proportion to their powers, a practice
" which led to much emulation and occasionally to fatal results.

    " As soon as all the first set of divers had come up, and their
" takings had been examined and thrown into the hold, the second
" set went down. After an interval, the first set dived again, and
" after them the second; and so on turn by turn. The work was
" very exhausting, and the strongest could not dive oftener than
" seven or eight times in a day, so that the day's diving was
" finished always before noon.

    " The diving over, the vessels returned to the coast and dis-
" charged their cargoes; and the oysters were all thrown into a
" kind of park, and left for two or three days, at the end of which
" they opened and disclosed their treasures. The pearls, having
" been extracted from the shells, and carefully washed, were placed
" in a metal receptacle containing some five or six colanders of
" graduated sizes, which were fitted one into another so as to leave
" a space between the bottoms of every two, and were pierced with

---

* The writer is obviously in error when he states that the diving stone is "tied" to
the diver's foot, that a diver cannot dive oftener than seven or eight times a day and also
in his account of the method of extracting the pearls from the decaying oysters.

His statement that one day's catch (not necessarily the first) belongs expressly to the
King (Nayakkan of Madura) or Setupati according to the locality where the fishery takes
place is correct only with regard to India. This privilege was frequently contended for
by the Nawab of the Carnatic at the Aripu fisheries but was consistently refused by the
Dutch, who however allowed the exaction to be made by mutual agreement between this
potentate and those of his subjects taking part in the fisheries held on the Ceylon side,
the suzerain right to one day's fishing (Valy or Wally) being reserved at Aripu by the
Dutch Government as one of its sources of revenue,

" holes of varying sizes, that which had the largest holes being the
" topmost colander, and that which had the smallest being the
" undermost. When dropped into colander No. 1, all but the very
" finest pearls fell through into No. 2, and most of them passed
" into Nos. 3, 4 and 5 ; whilst the smallest of all, the seeds were
" strained off into the receptacle at the bottom. When all had
" staid in their proper colanders, they were classified and valued
" accordingly. The largest, or those of the first class, were the
" most valuable, and it is expressly stated in the letter from which
" this information is extracted that the value of any given pearl
" was appraised almost exclusively with reference to its size, and
" was held to be affected but little by its shape and lustre. The
" valuation over, the Dutch generally bought the finest pearls.
" They considered that they had a right of pre-emption. At
" the same time they did not compel individuals to sell, if un-
" willing. All the pearls taken on the first day belonged by
" express reservation to the King or to the Setupati according as
" the place of their taking lay off the coasts of the one or the other.
" The Dutch did not, as was often asserted, claim the pearls taken
" on the second day. They had other and more certain modes of
" making profit, of which the very best was to bring plenty of cash
" into a market where cash was not very plentiful, and so enable
" themselves to purchase at very easy prices. The amount of
" oysters found in different years varied infinitely. Some years
" the divers had only to pick up as fast as they were able, and as
" long as they could keep under water, in others they could only
" find a few here and there. In 1700 the testing was most encour-
" aging, and an unusually large number of boat-owners took out
" licences to fish ; but the season proved most disastrous. Only a
" few thousands were taken on the first day by all the divers
" together, and a day or two afterwards not a single oyster could
" be found. It was supposed by many that strong under-currents
" had suddenly set in owing to some unknown cause. Whatever
" the cause the results of the failure were most ruinous. Several
" merchants had advanced large sums of money to the boat-owners
" on speculation, which were, of course, lost. The boat-owners
" had in like manner advanced money to the divers and others, and
" they also lost their money." *

* Thurston, E.   " Pearl and Chank Fisheries of the Gulf of Manaar," *Madras Museum Bulletin* No. 1, page 9, Madras, 1894.

The fishery of 1708 appears the next that was held, and one that gave a satisfactory return. At this fishery 398 free stones were allowed as follows :—

"*List of free stones according to ancient customs.*

96½ to the Naick of Madura—4 Xtian, 92½ Moorish.

60   to Theuver—60 Moorish.

10   to Head Moorman of Cailpatnam—5 Xtian, 5 Moorish.

185   to the Pattangatyns of this coast—all Xtian stones.

30   to those of Mannar

13   to those of Jaffnapatam.

3½ lost by 4 Moors who died in the fishery.
——
398   Stones free, valued at Pards (Pardãos) . . 3,591."

The 185 stones given to the Pattangatyns or headmen of the Paravas were in the nature of remuneration to these men for assistance in inspecting the banks, in guarding any oyster banks discovered, in recruiting divers and in superintending operations during the course of the fishery. All these stones are specifically termed "Christian stones," meaning that the divers using them were Christians (Paravas), whereas those allowed to the Nayak and Tēvar (Setupati of Ramnad) were all Moorish, save for four Christians. The explanation of this division is that the two great Muhammadan settlements, Kayalpattanam and Kilakarai were situated respectively in the territories of the Nayak and the Setupati, whereas the sovereignty over the Christian Paravas was vested expressly, first in the Portuguese and then by conquest in the Dutch.*

---

* The employment of native headmen in the examination of the Pearl banks and in the management of the fisheries and their remuneration by the grant of similar privileges to the above were continued by the Ceylon Government up to 1863   Latterly there were but five employed, namely the Adigar of Mannar the Maniagar of Karaiyur, two Adapannars and a Pattankoddi   The remuneration was five stones (the equivalent of one boat) to the Adigar, two to the Maniagar and one apiece to the other headmen   Sir William Twynam (Report on the Ceylon Pearl Fisheries, 1900) states that they generally sold their privileges at the beginning of the fishery   The Adigar of Mannar had the last privilege in 1863   Regarding the quality of the assistance they rendered Sir William Twynam remarks   " I found them highly intelligent and well-informed men, well acquainted with the Pearl banks and matters connected with the fisheries   They helped Mr Vane and Captain Pritchard to carry on the fisheries in 1855, 1856, 1857, 1858, 1859, and 1860 and the Adigar of Mannar rendered me very valuable service during the fishery of 1863." With this favourable estimate I heartily coincide both in regard to the present Adigar of Musali and the present representatives of the fishing and diving communities on the Indian coast ; a tactful and sympathetic attitude and the avoidance of any act or speech likely to arouse their prejudices soon win their confidence and in the present condition of the management of the Indian Banks the assistance they can render is not in any way to be despised.

The financial account of this fishery which has fortunately survived the vicissitudes of time is dated 24th May 1708 and is as follows :—

"TUTICORIN, 24th May 1708

In this fishery the whole number of stones employed was		4,321½
	P'DS.	
Namely, 2,380 Xtian at 7 pardaws each of 10 fanams	16,660	
1,551½ Moorish at 12 pardaws each of 10 fanams	18,618	
390 Heathen at 9½ pardaws each of 10 fanams	3,705	
Pards.	38,983	
Deduct 398 free stones at above value	3.591	398
Pards. ...	35,392	Sts. 3,923½
About ... fl.	106,176	
= £9,000 "		

From this statement we learn that the total of stones employed reached the astonishingly large number of 4,321½, considerably more than the number of divers who attended the Ceylon Fishery of 1903,[*] a fishery which gave the prodigious total of 41,169,637 fished oysters and a Government gross revenue of Rs. 8,30,000

Are we to infer that this Tuticorin Fishery of 1708 although yielding but £9,000 (Rs. 1,35,000 at the present exchange) to the Government was not so productive of oysters as the Ceylon one instanced ? In the absence of other particulars we have no means of judging with certainty but as the average price per stone is some 9 pardãos, each equal to about four guineas, and as this sum represents the license to fish accorded to a diver for the whole period of the fishery we may infer with some degree of probability from the large number of men engaged that the total catch may have been equally large. Under the conditions that rule at the present day, Government obtains a greater profit upon the

---

[*] The total of divers who attended the Ceylon Pearl Fishery of 1903, was 3,922.

fisheries, receiving two-thirds of the entire catch; hence the
receipts from the sale of the right to fish probably made the
fishery much more profitable to the subject in the early days of
Dutch rule and acknowledged receipts of fl. 106,176 (Rs. 1,35,000)
would represent a fishery on a scale of magnitude comparing most
favourably with the fisheries held during the last half of the 19th
century.

The proportions of divers supplied from the three religions then
prevalent is also shown by this account, namely—

     4,760 Christians (Paravas).

     3,103 Moormen.

      780 "Heathen", i.e., Hindus.

These 780 Hindu divers did not represent a remnant of Paravas
remaining unconverted to Roman Catholicism, but belonged
probably to the Kadeiyar caste of lime gatherers and burners
from which caste the ranks of the divers are in part recruited at
the present day.*

So far I can ascertain scarcely any divers practising the Hindu
religion have attended any pearl fisheries during the last half
century, while the relative proportionate strength of Christians and
Muhammadans has gradually tended to the preponderance of the
latter, so that at recent fisheries the Muhammadans outnumber the
Christians, an increase due partly to larger families reared by
the former and to the more regular and abstemious lives they
lead.

In my report on the Ceylon fishery of 1904 I noted the marked
superiority of Muhammadan over Christian divers in the number
of seconds they remain under water and in the greater number of
oysters collected per dive—a superiority that makes the work of
the former more productive and valuable. This appears to have
been recognized in a very practical manner in the old fisheries we
are now considering, as we see from the account given above that
the Moorish stones sold for 5 pardãos more than those of the
Christians, the rate of market valuation being as 12 to 7. Strangely
enough the Hindu stones occupied an intermediate value, being
9½ pardãos per stone.†

---

* Many of this caste are now converted to Roman Catholicism, and it is from this
division of the caste that the present supply of Kadeiyar divers is drawn.

† Probably the differences were discriminatory, penalizing the Muhammadan and
the Hindu for their religious beliefs.

At the Ceylon fishery of 1694, a similar disparity is noticeable there being then sold :—

    1,290 Christian stones at 6½ Rixdollars

    204 Gentoo (Hindu) at 9   do.

    1,268 Moorish   at 11½   do

In the accounts of this last fishery, I notice an entry of " 13¾ ammonams of concealed arrack found in the bushes and out of the way places, sold at 6 Rixdollars the ammona = 82½ Rixdollars," from which I fear we have to infer that the Paravas were as greatly addicted to this indulgence 200 years ago as they are now and also in the days of St Francis Xavier !

No further fishery appears to have taken place between 1708 and the date of the relinquishment of the Governorship of Ceylon by Baron Van Imhoff in 1740

Such a lengthy intermittence in productiveness had been a source of continual regret to his predecessors, so that when Van Imhoff wrote his valedictory memoir in 1740 for the information and guidance of his successor, he urged a departure from the policy of conducting the fishery in aumany or directly on behalf of Government Under such conditions the Government had made its profit by the sale of the right to use diving stones at the fishery (amounting virtually to the taxation of the divers employed) and of the sundry duties levied in the fishing camp (" Exchange," Bazaar and cloths).

This memoir is perhaps the most valuable and statesmanlike administrative record left by any Dutch Governor Therefore it is well if I give *verbatim* the portion relating to the condition of the Pearl Banks at this time and to the alternative policy of administration advocated here for the first time

Extract from the Memoir of His Excellency G W Baron Van Imhoff on his departure from the Government of Ceylon, left for the instruction of his successor, His Excellency William Maurits Bruynink.

" The Aripo and Tutucoryn Pearl Fisheries are certainly to be " reckoned among the sources of produce to this island or rather of " revenue, for the profit which the Company derives from the holding of a " fishery must rather be classed under the latter head than under the " former, as it consists in different duties which are paid for diving those " banks, and divers sums paid for the stones used in catching the oysters, " and some part in oysters themselves paid as taxes, which are sold when

" the fishery is at an end ; duties are also paid on what is called the
" exchange, and on cloths which are brought to the bazaar, but the
" Company does not in fact obtain any pearls, nor is there even a chance
" for the Company to purchase any pearls there, although the highest
" authorities have so often endeavoured to do so, for at the fishery pearls
" are sold at so high a price that the Moors are cunning enough to rub up
" even old pearls and to bring them there for sale, with a certainty of
" taking in the unwary, and deriving more profit from it.

 " But it is not so much a matter of concern whether the fishery is to
" be called a source of revenue or of produce, as whether it can in reality
" be looked upon as a source of advantage and profit derived from Ceylon,
" or whether it is more glitter than gold, as many things are which belong
" to the Company, which shine uncommonly, but have no real substance
" This question is neither a novel nor unfounded one, and to properly
" answer it, we must weigh against the advantages which we have just
" detailed, the inconveniences, discomforts, noise, expenses the risks of
" the Commissioners, the employment of the Militia, the consumption of
" provisions, the dangers of ships, etc ; we must also mention the hazards
" run by a few hundred men sent to keep immense crowds in order, and
" their exposure to sickness and death as well after the fishery as during
" its continuance from the stench of the oysters ; the price of provisions is
" also enormously increased ; the Company's trade in cloth is discontinued
" for a long time from the prevalence of smuggling which is occasioned by
" the immense numbers of persons resorting to those parts of the island ; we
" may also add that pepper is smuggled away, as well as arecanuts, although
" it would be thought that a multitude of more than 100,000 persons who
" consume these nuts for the space of two or three months should give
" some profit, yet the Company draws nothing from it   If therefore all
" these matters be weighed, one against the other, it must be decided, as I,
" for my part, maintain, that unless the fishery be indeed a full and opulent
" one, all others must be prejudicial to the Company's interests , and it
" were really desirable that no such fisheries should take place, but that
" there should be an annual rent for the diving of the banks, as now takes
" place with regard to the chanks, with a limited number of persons and
" of boats , or in some other convenient way that a mode should be devised
" to acquire for the Company the profit which they should derive from the
" fisheries, both here and on the opposite coast, as Lords of the country,
" without the holding of any public fishery   The bad condition of the
" pearl-banks on both sides the coast has lasted for some years, and there
" is now no prospect of an early fishery ; yet this cannot be attributed to

"any disorder in the country any more than the want of purchasers for
"our Madura cloths.   This is mere chance and experience has shewn that
"the banks have lain fallow for a much longer time than has as yet been
"the case on this occasion, and it is useless therefore to seek for the causes
"of things which are neither uncommon nor unheard of. I only mention
"this cursorily, and to prove that the interest of the Company requires
"that an examination of the banks should take place every two years, if
"not every year, which indeed is not absolutely necessary, and the expense
"may therefore be spared, yet from time to time or say every two years,
"an investigation should take place

      "As far as the inspection of the Aripo banks is concerned, Tutucoryn
"boats should be employed, as their Honors recommended in a recent
"despatch, a Dissave from Jaffnapatam should also be present to see that
"no neglect takes place, and to summon the boats from the opposite coast,
"and to take all necessary precautions when appearances prove favourable,
"in order to ensure a good result ; for the Company has been shamefully
"treated in this respect since the fishery of 1732.   Indeed there are many
"natives who pretend to give reasons for the failure of the banks, and who
"say that the multitude of persons forced there against their will have
"ruined the banks, whilst others looked to their own profit too much, and
"also that the divers have not spared the young oysters, and that this
"accounts for the nakedness of the banks which have not yet recovered
"from their last pillaging. All this is as probable as the pretexts of the
"country being under a spell, but to end this matter we will pray God that
"the Island may never again suffer losses such as it now sustains from one
"cause or the other."

Four years after the date of this memoir, namely in 1744, Baron
Van Imhoff became Governor-General of the Netherlands Indies
and immediately called attention to his Ceylon memoir He
desired to be informed whether it would not be advisable to dis-
continue open fisheries and preferable to rent them out to a
single individual. Van Gollenesse, the new Governor of Ceylon,
in a closely-reasoned reply, * meets and refutes the objections
likely to be raised to this change and strongly advocates an
alteration in the method of conducting the fisheries, which was
thereupon sanctioned

Accordingly the fishery which took place in 1746 was conducted on this new footing all free boats and stones were

---

* See Appendix D, pages 179—183.

abolished, for, as the Governor states, these privilges were merely conceded to the Nayak and the Setupati because the greater number of the dhonies and people required at a public fishery came out of their territory, and these would not be necessary if the diving took place with a limited number of persons. All privileges were therefore withdrawn and the following instructions were given to the Commissioners of the Pearl Fishery by Governor Van Gollenesse:—

"If it should happen (which is however improbable) that the "Nabob or Theuver should send their Ambassadors to be present "at this Fishery, and to take care of their pretended rights, they "are not to be allowed to land, but some armed boats well filled "with men and ammunition must be sent to meet them, and they "must first be warned in a friendly way to depart, and if this be "ineffectual, the matter must be treated more seriously, and you "must order the Commanders of our boats that they are by no "means to permit any armed foreign boats of a suspicious kind to "come within range of their shot, and if warning given does not "turn them away, they must fire on them at once."

1747 This year the unproductive cycle that had prevailed so long on the Tuticorin side was broken, the fishery being *rented out* for 60,000 florins (£5,000).

The change of system now introduced involved the abolition of the privilege of free or untaxed divers hitherto granted to the native rulers of Madura and Rāmnād, whose dominions had now merged in those of the Nawab of the Carnatic who had dispossessed the last of the Nayak dynasty of Madura in 1736. He, a more powerful ruler than the Nayak, did not acquiesce without stout opposition and at the 1747 fishery it had been thought wise to permit the renter of the fishery to give 30 free divers to the Nawab for which concession the renter received a proportionate reduction in the stipulated price of his rent.

Two other fisheries, also rented out, took place in the two next succeeding years. The rental of that of 1748 amounted to 114,720 florins (£9,560), while that of 1749 was florins 63,600 (£5,300).

At the former, 35 free divers were again allowed to the Carnatic overlord, much against the Dutch Company's will however, for we read in the secret instructions sent to the "Company's

Commissioners in the rented fishery on the coast of Madura " before the fishery of 1749 :—

"We think it necessary to inform you also that as the Armane "may cause much injury to our defenceless Linen Factories, we "granted the Nabob, on his urgent request, 35 divers in the two "last fisheries, but Their Excellencies (the Government of Batavia) "did not approve of this concession, and therefore, in case His "Envoys should again claim this grant from you, you must "endeavour to reduce the number to 17 or 18 divers, showing that "even this will give greater profit than the 96½ which the Nabob "had formerly in an open fishery, when the whole number of "stones amounted to several thousands.

"But if you cannot effect this, and if you see any risk for the "Company's becoming embroiled with the Regent by a pertina-"ccous refusal, you will then be empowered to grant 30 or 35 divers, "but it must appear to be done without our knowledge, and on "your own private authority.

"But if the Catta Theuver, or any other native chief, should "request a similar concession, you must refuse it flatly."

Between 1749 and 1784 I can trace no record of any further pearl fishery off the Tuticorin coast save the suggestion of one in 1771 furnished by the existence of a set of "Conditions of a rent of the Tutucoreen Fishery of 1771." Article XXX of these conditions reads—

"Lastly the renter of the fishery must admit 20 dhonies of the "Armanie or Regent of Madura, with 96½ stones and two dhonies "on the part of the Catte Theuvers,* manned in the same manner "as the Renter's dhonies, which 22 dhonies, together with 180 of "the renters, shall fish throughout the whole fishery without the "renters being permitted to make any demand on that account."

Of the fishery of 1784, the only particulars I have (furnished by the Madras Government) are that it was held on the Tolayiram Par, giving to the Company, which fished it departmentally, a

---

* The Tevar, the Setupati Raja of Rāmnād, made a treaty with the Dutch in 1767, whereby in exchange for the possession of Pamban Pass and the surrender to the Dutch of the right to levy dues on shipping passing therethrough, the Dutch agreed to grant the Setupati two free diving boats at all the future pearl fisheries held on the banks lying off the Coast of Madura, or "under the territory of Tutucoryn together with the privilege *to purchase* from the Dutch Government in every fishery held on the Ceylon side five boats at the same price as the renter should contract."

gross revenue of 20,000 cully chucrums. This at Rs. 2 1-11½ per cully chucrum gives a return in rupees of Rs. 42,447-14-8 Caldwell (*loc cit*) notes that this fishery was the first conducted by the English East India Company, Tuticorin being then temporarily in their hands.

This long series of blank years extending, with the doubtful exception abovementioned, over a period of thirty-five years, may have been due up to 1768 to imperfect inspection or to natural causes or to a combination of the two, but the intermittence thereafter arose in the main from the reluctance of the Dutch to agree to the pressing demands of the Nawab of the Carnatic to participate upon exorbitant conditions in the profits arising from the fisheries both on the Ceylon and the Indian Banks. As already mentioned, at the Ceylon Fishery held at Aripu in 1768 violent disputes occurred with the Nawab's envoys who went to the fishery attended by a large body of armed sepoys and tried to carry matters with a high hand

The Dutch loth, with their usual caution and fear for the interruption of their cloth monopoly in Madura, to bring matters to a crisis, preferred to let the pearl fisheries remain virtually in abeyance till a settlement was effected on equitable terms which meant the curtailment if possible of the Nawab's pretensions

It was not till 1786 that the Dutch, pressed by the English Government in Madras (to whom the Nawab had appealed as his ally and virtual suzerain) to effect a settlement of the long-standing dispute, made provisional terms with the Nawab * By these the Nawab obtained much greater advantage than had been contemplated twenty years previously, due to the dwindling power of the Dutch and their growing fear of the rapid extension of the military power and commercial supremacy of the English East India Company.

The chief articles of the agreement affecting the pearl fisheries were that the Nawab should be granted one-half of the profits arising from fisheries off the Madura coast and have 36 free dhonies at any fisheries held on the Ceylon side, privileges allowed in return for a confirmation of the Dutch trading monopoly in Madura cloth—ever one of the most lucrative sources of revenue to the Dutch Company.

---

* Two years later a definite treaty on the same lines was signed by the Dutch.

6

The treaty, however, was never fully ratified, but by the advice of the English Governor of Madras its terms were allowed to govern the fisheries of 1787 and 1792. the profits therefrom being accordingly shared equally by the Sircar or Government of the Carnatic (i.e., the Nawab) and the Dutch Company.

The fishery of 1787 took place on the Tolayiram Pär and gave a gross revenue of Rs. 63,000 : that of 1792 upon the Uti, Uduruvi, Kilati and Attuveiarpagam Pärs, which lie inshore of the Tolayiram Pär, it yielded Rs. 42,525 to the joint Governments.

Except with regard to the conduct of these two fisheries, the treaty never came into force, the Madras Government steadily refusing its consent because of the objectionable clause relating to the cloth monopoly In this unsettled condition, marked by the continual interchange of despatches between Colombo and Madras, matters remained till the Dutch dominion of the Pearl Banks on both sides passed to the British in 1796.

(d) THE PEARL BANKS UNDER THE BRITISH.

Not long after the acquisition of the Pearl Banks by the British, the districts bordering the coast in this region and now known as that of Tinnevelly in the south and Rämnäd in the north, passed to the British from the Carnatic Nawab Thus the "Lords of the Pearl Fishery" acquired sovereign rights over the districts supplying the whole body of divers and by their own power could ensure safe conducts from Madura Rämnäd, Bombay, and Madras to the dealers in pearls whose attendance is necessary to the success of any fishery. The dues levied for assistance by local potentates, the source of constant anxiety and loss to the Portuguese and the Dutch, were brought to an end and for the last century we hear of no privileges allowed save a few on a reduced scale to the headman of the Paravas It is noteworthy to observe that this system of remuneration by fishery privileges of which the last remaining trace was abrogated in Ceylon in 1863, still lingers in the management of the Indian banks, the headman of the Paravas having the right to employ a limited and specified number of "free" boats at each fishery in return for help rendered during the inspection of the banks and at the fisheries when held.

This hereditary chief or Jathi Talaivan * of the Paravas, like many of the descendants of natives of Ceylon who gave assistance to the Portuguese, bears the honorific prefix of " Don," while the name of the late holder of the Chieftainship—Gabriel de Cruz Lazarus Motha Vaz—further indicates the intimacy of his family's connexion with Portuguese rule.

His duties consist in accompanying the Inspector on his periodical visits to the banks—a duty formerly performed directly by the headmen themselves,—in furnishing guards to the banks to be fished, in supplying Government with information of any accidental finds of oysters by fishermen and in acting as intermediary between the Government and his caste with a view, by the exercise of his influence, to ensure the attendance at a fishery of an adequate supply of boats and divers.

In 1889, the Madras Government recorded their appreciation of the assistance rendered by the Jathi Talaivan and " directed that his privilege of being allowed the take of two boats be continued."† Each boat is understood to carry ten divers.

Subsequently, in 1891, the Madras Government, while confirming the general principle of privilege remuneration to the headman named, adopted the more satisfactory regulation of placing the extent of the remuneration upon the basis of a sliding scale, allowing him but one boat when the Government boats numbered 30 or less, two for 31 to 60 boats, three for 61 to 90 boats employed and so on in this ratio

The value of the Jathi Talaivan's two privilege boats in the 1890 fishery was Rs. 1,424, in that of 1900 only Rs. 872.

During the 100 years ending 1922 there appear to have been the following fourteen fisheries, viz —

---

* Literally " Head of the Caste." His full title is Jathi Talaivaimore—the suffix " more" being, I believe, an honorific addition made by the Portuguese The Tamil form in simply Jathi Talaivan. Caldwell (*loc. cit.*, p 147) notes that in 1783, Tuticorin being temporarily in possession of the English, the Receiver of the revenues of the Tinnevelly country was instructed by the authorities " to present an honorary dress to the head of the Paravas (the Jati Talaivar) in the name of the Madras Government."

† Proceedings of the Board of Revenue, Madras, No 702, 1889. The amount which this privilege realized to this headman at the 1889 fishery was Rs. **7,620.**

(e) TABULATION OF THE LOCALITIES AND THE FINANCIAL RESULTS OF ALL INDIAN FISHERIES FROM 1800 to 1922.

Year	Number of working days	Number of boat fishings	Average number of boat fishings per day	Total number of oysters lifted on Government account	Grand total of oysters fished	Age of oysters	Gross Government revenue	Net Government revenue	Average price per 1,000	Locality	Authority
(1)	(2)	(3)	(4)	(5)	(6)	(7)	(8) RS.	(9) RS	(10) RS A P	(11)	(12)
1805 1807 1810	.. ..	..	..	22,096 658 plus oysters fished by Government departmentally	71,647,305	..	39,109 2,91,539 2,38,897	..	..	Velangu Karuwal Pār Tolayiram Pār Do	From particulars furnished by the Madras Government, supplemented in the cases of the 1805-15 Fisheries by the Jathi Talavan's records
1815	.	...	...			.	Nil (an unsuccessful fishery)	...	.	Velangu karuwal, karai Karuwal and Tiruchendūr Puntotta Pārs	
1818	.			..	More than 15,500,000		1,67,708			Saith Kudamuttu, Kudamuttu, Pudu Pārs	
1822 1828	..	..		..	7,541,940	.	1,55,593 70,127		.	Saith Kudamuttu, Pudu, Kudamuttu, kadian, kanavai and Rajavukku Sippi Sonchcha Pārs	

Year								Locality	Reference
1830	.	.		...	12,858,993	...	1,01,639		Tolayiram, Utu, Uduruvi, Kilatu and Paduta Marikan Tundu Pārs	Proceedings, Board of Revenue, No. 484, 1889.
1860-61	:	.		:	...	:	2,50,276	2,21,801		
1862	...	.		...	...	.	1,29,003	1,10,619	Cruvian, Cruvian Tundu, Nagara, Utu, Uduruvi, Atonbadu, Kilati, Devi, Pernandu and Vaippai Karai Pārs / Kural Karuwal and Velangu Karuwal Pārs	
1889	52	1,731	33	8,423,925	12,600,531	5 years.	1,83,984	1,38,453 22 8 6	Tolayiram Par	
1890	40	857	22	1,224,816	1,866,762	6 ,, about 4 years	25,061	7,803 20 10 0	Do	Proceedings, Board of Revenue, 1900, No 208
1900	15	746	50	1,873,426	2,810,136		19,461	11,033 10 4 9	Teradi Puli Piduta and Tundu Pāis.	
1908	20	528	26	734,428	1,101,642	4 years	10,218	7,282 13 13 1	Tolayiram Par	G O No 1946, Rev Dept, 16th July 1908.
1914	20	562	28	355,611	533,416	about 3½ years	16,542	2,497 26 6 4	Off Tondi, Palk Bay	"Madras Fisheries Bulletin" Vol. 8, Part IV

The discovery of beds of pearl oysters in Palk Bay, was the result of an inspection made by the writer in 1914 The discovery and the details of the fishery off Tondi that followed are described in Bulletin No. 8, part IV. The oysters lay almost entirely between the 5½ and 6 fathom lines, on very muddy sand, attached in the main to dead shells Another small bed is found north-east from Rameswaram town in about the same depth. These beds, unlike those in the Gulf of Mannar, appear to be persistent, for scattered oysters can always be found there It is probable that these permanent deposits give rise to the initial spat-falls that are requisite to replenish the Mannar beds after these have become depopulated.

The oysters are at the present day fished on Government account and according to the arrangements in force at the last three fisheries the Government claim two-thirds of the total catch, selling their share at auction. The remaining third is the remuneration allowed to the divers and boat owners.

In the following table I list and contrast all the fisheries of which I can find record on the Tuticorin and Ceylon sides, respectively, of the Gulf of Mannar:

Particulars of Pearl Fisheries from 1658 to 1922.

Off Madura coast.			Off Ceylon coast.		
Year of fishery.	Government proceeds.	Remarks.	Year of fishery	Government proceeds	Remarks
				£ s d	
1663	Fl. 18,000	.....	...	...	...
..		......	1666	4,913 15 11	Fished on Government account giving net profit as shown.
			1667	6,160 7 5	Do.
1669	...	} I can find no particulars	..		
1691	...	} of profits realized.	..		
...	..	... ..	1694	5,264 16 1	Do.
...	..	..	1695	6,177 3 9	Do.
...	...		1696	6,331 18 4	Do.
..		......,	1697	6,453 0 0	Do.
1700	Very meagre.	A disastrous 3 days' fishery	..	..	
1708	£ 9,000 ...	= Fl. 1,06,176 gross proceeds.	1708	8,848 0 0	Do
...	..	.. ..	1732	Not ascertainable.	An unproductive fishery
..	....	..	1746	4,766 13 4*	Fishery rented out to adventurers.
1747	£ 5,000 .	= Fl. 60,000	1747	21,400 0 0	Do
1748	£ 9,560 ...	= Fl. 1,14.720	1748	38,580 0 0	Do.
1749	£ 5,300 ..	= Fl. 63,600	1749	68,375 0 0	Do
...	...	... .	1750	5,940 0 0	Fishery of 6 days only.

{ Fide Governor Schreuder's Memoir of 1762, as given in Lee's edition of Ribeyro's "Ceylon."

* According to Schreuder ; £ 12,000 according to Lee (Ribeyro).

Particulars of Pearl Fisheries from 1658 to 1922—cont

Off Madura coast.			Off Ceylon coast.		
Year of fishery.	Government proceeds	Remarks	Year of fishery	Government proceeds.	Remarks
	Rupees			£　s.　d.	
..	..	....	1753	6,360　0　0	Fishery rented out
..	..		1754	1,469　0　0	Do.
.		No fisheries held between 1768 and 1784 owing to disputes with the Nawab of the Carnatic.	1768	Not ascertainable	Very unsuccessful on account of bad weather.
1784	42,477 Gross	−20,000 cully chucrum. Fished departmentally.	..	..	..
1787	63,000 Gross.	⎰Rent received in equal shares by the Dutch and the Nawab of the Carnatic.	..	..	...
1792	42,525 Gross.	⎱	.	..	
...	...	.. ...	1796	37,096 15　0	Fishery rented out.
..	..	..	1797	123,982 10　0	Do.
...	...	...	1798	142,780 10　0	Do.
.	..	......	1799	23,319　7　6	Net proceeds— fished on Government account
...	...	.....	1801	**Rupees.** 1,50,227	Gross proceeds—fished on Government account.
	.	......	1803	1,63,154	Do
	...		1804	7,20,202	Gross proceeds
1805	39,109	Gross proceeds	1806	4,12,842	Gross proceeds.
1807	2,91,539	Gross proceeds .			
1810	2,38,897	Gross proceeds ...	1808	8,42,577	Gross proceeds.
			1809	2,72,463	Do.
			1814	10,51,876	Do.
1815	Nil.	An unsuccessful fishery, giving no revenue	1815	5,842	Do
1818	1,69,708	..	1816	9,266	Do
1822	1,55,693	Gross proceeds	1820	30,410	Gross proceeds.
1828	70,127	Do.	1828	3,05,234	Gross proceeds.
...	...		1829	3,82,737	Do.
1830	1,01,639	Gross proceeds ..	1830	2,22,564	Do
...	..	. ..	1831	2,93,366	Do.
...	...	.....	1832	45,810	Do.
	.	...	1833	3,20,896	Do.
	...	..	1835	4,03,460	Do.
	..	.. ..	1836	2,54,935	Do.
...	..	......	1837	1,06,312	Do.
...	.	.. ....	1855	1,09,220	Do.
..	..	. ...	1857	2,03,633	Do.
...	..	.....	1858	2,41,200	Do.
.	...	.	1859	4,82,159	Do.

Particulars of Pearl Fisheries from 1658 to 1922—cont

Off Madura Coast.			Off Ceylon Coast		
Year of fishery	Government proceeds	Remarks	Year of fishery.	Government proceeds	Remarks
	Rupees			Rupees	
1860 1861 }	} 2,50,276	Gross proceeds	1860	3,65,816	Gross proceeds
1862	1,29,003	Do.	...		...
	..		1863	5,10,178	Gross proceeds.
..	.. .		1874	1,01,199	Do.
	.		1877	1,89,011	Do
.	..		1879	95,694	Do
...			1880	2,00,152	Do
...	..	...	1881	5 99,533	Do
...			1884	17,153	Do
..	.. .	.	1887	3,96,094	Do
			1888	8,04,247	Do
1889	1,89,984	Gross proceeds	1889	4,98 377	Do
1890	25,061	Do.	1890	3,13,177	Do.
			1891	9,63,748	Do
1900	19,461	Gross proceeds			
			1903	8,29,348	Gross proceeds
...	..	...	1904	10,65,752	Do
.	.		1905	25,10,727	Do
...	..	.. ..	1906	3,10,000	Nett, Rented out
.			1907	3,10,000	Do.
1908	10,218	Gross proceeds	...	....	.. .

Twenty five fisheries within 264 years   Sixty-one fisheries within 264 years

Note—The particulars regarding the Ceylon fisheries prior to 1801, are taken from the appendix to Lee s translation of Ribeyro's " History of Ceylon " Colombo, 1847   The later figures are from official returns furnished to me by the Government of Ceylon.

The list of fisheries held on the Ceylon side is, I believe, exhaustive, possibly two or three may be omitted from the enumeration of those upon the Indian coast   As it stands we have a total of twenty-five fisheries recorded from the latter locality as against sixty-one from the former during the period of 264 years from 1658 to 1922.

Comparing the two lists a noteworthy feature is that many of the fisheries held on the Tuticorin banks coincide with blank years on the Ceylon banks

Thus out of all the Indian fisheries, those of 1663, 1669, 1691, 1700, 1784, 1787, 1792, 1805, 1807, 1810, 1818, 1822, 1861, 1862, 1900 and 1908, sixteen out of the total of twenty-five, were years in which no fishery took place on the Ceylon side.

Again this fact may be correlated to two others—

(a) that usually any particular Indian fishery was preceded at a distance of from two to three years by a Ceylon fishery and

(b) that in the same way each Indian fishery was followed at a similar interval by one on the Ceylon banks

Thus the Indian fisheries of—	were preceded respectively by Ceylon fisheries held in—
1669	1666, 1667
1700	1695, 1697
1747—1749	1746
1805	1799, 1803 and 1804
1807	1804, 1806
1810	1806, 1808.
1815	1809, 1814.
1818	1814—1816.
1822	1820
1830	1828, 1829
1860—1862	1857—1860
1889, 1890	1884, 1887
1908	1903—1907.

while in the same way the Indian fisheries of—	were followed respectively by Ceylon fisheries in—
1663	1666, 1667
1691	1694—1697
1747—1749	1750, 1753 and 1754.
1792	1796
1805	1806, 1808
1807 and 1810	1809, 1814
1815	1816, 1820
1818	1820
1822	1828
1828	1829—1833.
1830	1831—1833 and 1835
1860—1862	1863
1900	1903—1907

Such regularity of alternative succession extending over 75 per cent of the fisheries held on the Indian side appears to be more than a mere coincidence and lends weight to an opinion that has gradually been taking shape and developing in my mind that the beds on the opposite sides of the Gulf confer reciprocal benefits upon one another and that the Ceylon banks are frequently replenished from those off the Madura coast and, conversely, that the latter obtain most of their deposits of spat from the Ceylon side

7

The following table furnishes the valuation particulars of a sample of oysters lifted from a Tuticorin bank :—

Statement of the valuation and produce of 8,500 oysters lifted from the Teradipulipiditta Pär in October 1899.

Description	Size in sieve.	Number.	Quantity in chevu	Weight Kalanji.	Weight Manjady	Total Kalanji.	Total Manjady	Total value	Per chevu	Per kalanji.
									STAR PGDS.	STAR PGDS.
								PS A. P		
Ani	50	1	$\frac{3}{320}$	...	$\frac{3}{16}$	$\frac{1}{10}$	...	7 0 0	80	.
	80	1	$\frac{3}{320}$	...	$\frac{6}{16}$	$\frac{1}{32}$	...	4 6 0	80	.
Anatari .	80	3	$\frac{20}{320}$	...	$\frac{1}{2}$	...	$\frac{1}{2}$	9 10 0	45	...
Kuruwal .	20	2	...	...	$\frac{1}{4}$	...	$\frac{1}{4}$	7 14 0	.	60
Kalippu .	20	1	$\Big\}\frac{64}{320}$	...	$\big[\frac{1}{2}$	...	$\frac{1}{2}$	12 0 0	20	..
Podi Kalippu	80	1		...	$\big[\frac{1}{8}$	...	$\frac{1}{8}$			
Pisal . .	...	...	.	...	$1\frac{1}{2}$	...	$1\frac{1}{2}$	0 6 0	..	$1\frac{1}{2}$
Vadivu	100 / 200 / 400 / 600	...	...	...	3	...	3	31 8 0	...	60
Tul	800 / 1000	...	...	...	$6\frac{1}{2}$	...	$6\frac{1}{2}$	9 12 0	...	$8\frac{1}{2}$
Masitul ...	...	..	...	..	$5\frac{3}{4}$	...	$5\frac{3}{4}$	2 14 0	.	...
Shells pearls ..	.	...	...		$6\frac{1}{2}$	...	$6\frac{1}{2}$	0 12 0		
Total	...	..	..	..	..	...		86 2 0		.
Average per 1,000 oysters	...	...	...	...	...	.	..	10 2 0		.

IV—NOTE ON THE TOPOGRAPHY OF THE BANKS.

The charted pearl banks along the Indian coast of the Gulf of Mannar represent all those patches of rocky ground lying within the 10-fathom line known to the fishermen of that coast. Taken as a whole they deserve the name of pearl banks only so far as being so potentially. The number of these banks which have been known to bear mature oysters during the past century is limited to 23 at most, and except in the case of nine, none of them has been fished more than once in this long period. These potential pearl banks extend from Cape Comorin to Rāmēswaram island at the extreme head of the Gulf, a distance of over 100 miles. They consist of whatever rocky outcrops there are upon the surface of the wide sub-marine plateau which fringes the whole extent of this coast. This pearl bank plateau is widest in the south, in the neighbourhood of Cape Comorin, gradually narrowing as we proceed northwards. In the south it shelves to the 100-fathom line at an easy gradient, and everywhere the width of the plateau is considerably greater than anywhere in the pearl bank region on the Ceylon side.

The pars, as these banks are termed locally, may be arranged in three divisions—the Northern or Kilakarai, extending from Adam's bridge to Vaippar, the Central or Tuticorin, from Vaippar to the latitude of Manappad ; and the Southern or Comorin from thence southwards to Cape Comorin.

The central division is by far the most important , indeed so far as recent historical evidence goes the banks of this division are the only productive ones.

Many are extremely small , some have been described as having an area little greater than that of an ordinary sized room, and as they owe their separate entities to the detailed local knowledge of fishermen engaged not in pearling but in ordinary fishing it will conduce greatly to simplify the pearl bank management if, in future, the majority of these separate banks instead of being listed individually, be linked together into groups, the members of each group occupying adjacent positions and having similar physical and biological characteristics , some have characters rendering them entirely unsuited to the maturing of oysters and these may be deleted eventually once and for all.

The names of these banks arranged in order from north to south and classified into groups consonant with their relative geographical position and with their identity of physical and biological characteristics are as follows :—

A —Northern or Kilakarai Division

Composed of

I. Pamban Group	1.	Pamban Karai Par.
	2.	Pamban Velangu Par.
II. Musal Tivu Group	3.	Musal Tivu Par.
	4.	Solaka Karai Par.
III. Kilakarai Group	5.	Kilakarai Vellai Malai. Velangu Par.
	6.	Vellai Malai Karai Par.
	7.	Anna Par.

A.—*Northern or Kilakarai Division—cont.*

Composed of

IV. Tanni Tivu Group ... { 8. Nalla Tanni Tivu Par.
{ 9. Uppu Tanni Tivu Par.

V. Vembar Group .. { 10. Kumulam Par.
{ 11. Vembar Periya Par.

B—*Central or Tuticorin Division*

VI. Outer Vaippar Group ... 12. Vaippar Periya Par.

VII. Inner Vaippar Group { 13. Karai Par.
| 14 Devi Par.
{ 15. Pernandu Par.
| 16. Padutta Marikan Par.
| 17. Padutta Marikan Tundu Par.

VIII. Cruxian Group . { 18. Tuticorin Kuda Par.
| 19. Cruxian Par.
{ 20. Cruxian Tundu Par.
| 21. Vantivu Arupagam Par

IX. Uti Group { 22. Nagara Par.
| 23. Uti Par.
{ 24 Uduruvi Par.
| 25. Kilati Par
| 26. Attuvai Arupagam Par.
| 27. Attonpatu Par.

X. Pasi Par Group { 28. Pasi Par.
{ 28A. Pattarai Par.

XI. Tolayiram Par { 29. Kutadiar Par.
{ 30. Tolayiram Par.

XII. Puli Pundu Group { 31. Vada Onbadu Par.
| 31A. Saith Onbadu Par.
{ 32 Puli Pundu Par.
| 32A Kanna Puli Pundu Par.

XIII. Kanna Tivu Group { 34 Kanna Tivu Arupagam Par
{ and perhaps.
| 35 Tundu Par.

XIV. Nenjurichchan Group { 36 Nenjurichchan Par.
{ 36A. Par Kundanjan Par.
{ 36B. Mel Onbadu Par.

XV. Inner Kudamuttu Group. { 37. Pinnakayal Seltan Par.
{ 38. Sandamaram Piditta Par.

B.--Central or Tuticorin Division—cont.

Composed of

XV Inner Kudamuttu Group *—cont.*	39.	Irai Tivu Kudamuttu Pār
	39	Nadu Kudamutu Pār
	41	Kudamuttu Pār
	41A	Rajavukku Sippi Sotichcha Pār
	41B	Saith Kudamuttu Pār.
XVI. Outer Kudamuttu Group	40	Kovil Piditta Pattu Pār.
	40A	Sankuraiya Pattu Pār
	40B.	Nillan Kallu Pār
	40C.	Sattu Kuraiya Pattu Pār.
XVII. Kadeiyan Group	43	Kadeiyan Pār
	43A.	Kanawa Pār
	43B.	Pudu Pār
XVIII Karuwal Group	42	Naduvu Malai Piditta Pār.
	42A	Periya Malai Piditta Pār
	44	Karai Karuwal Pār
	45.	Velangu Karuwal Pār
XIX Odakarai Group	48.	Odakarai Pār
XX. Chodi Group	49	Chodi Pār
XXI Tundu Pār Group	46	Tundu Pār
XXII Manappad Group	47	Trichendur Puntoddam Pār
	50.	Sandamacoil Piditta Pār
	51.	Teradi Piditta Pār.
	52	Semman Path Pār.
	53.	Surukku Onbadu Pār.

C--Southern or Comorin Division.

XXIII Manappad Periya Pār

The remainder of the Southern banks cannot at present be grouped, as their positions are not marked on the Inspection Chart, and as I have had no opportunity to examine them   Their names will be found in the list on pages 61 and 62, numbered from 56 onwards

CENTRAL OR TUTICORIN GROUP.—The Central division corresponds both in latitude and in extent with the productive pearl bank region on the opposite Ceylon coast, Manappad point coinciding exactly with the latitude of the Ceylon Muttuvaratu Pār, while the Tolayiram Pār, off Tuticorin, similarly coincides, as well in latitude as in great relative extent, with the Cheval Pār, the bank of largest area and greatest productive importance on the Ceylon side.

GENERAL REMARKS.

The great majority of the pars are marshalled roughly in line parallel with and at a distance of from 7 to 8 miles from land From Kayalpattanam to Vaippar a second and outer series occurs, lying in rather deeper water. The depth of the inner series is in the main 7 to 8 fathoms, that of the outer 8 to 11 fathoms.

The *surface* of these pārs consists of a rock which appears in many instances to be of recent origin, rock formed by the consolidation of sand and dead corals *in situ*. The nature of the rock varies considerably, partaking usually largely of the present character of the circumjacent sand and as the latter on this portion of the Indian coast is made up principally of calcareous grains formed from the comminuted remains of shells and corals, so the calcrete is normally a more or less pure limestone. Occasionally the remains of corals are met with, and here and there the calcrete contains a varying amount of quartz sand. The proportion of quartz in no case is so great as that characterizing the typical quartzose calcrete so common on the Ceylon side. In several localities visited during the inspection, I am, however, of opinion that the exposed rock surfaces are not of contemporary origin, being of limestone too hard and compact to be considered a modern calcrete Further, where such latter calcrete does occur, it appears to me to form but a comparatively thin crust over the underlying more compact bed-rock of the plateau whereof the density and grain appeal to me as significantly identical with the extremely hard, compact rock forming the core of the Jafina islands and peninsula in the north of Ceylon, and of the rocky ridge that ends in Manappad Point on the Indian side

In no place did I see any shelly conglomerate, no rock in which the main constituent could be made out as formed from the accumulation of shells of pearl oysters, cockles and the like

It is impossible to say with any certainty whether the banks which appear at the present day to be the only banks from which we can reasonably expect to reap an occasional pearl harvest, have always had this character or whether the banks which were productive centuries ago and anterior to the advent of European control were situated further south. Certain facts and inferences incline me to suspect that the latter was, to some extent at least,

the case. There seems to be considerable evidence pointing to a
considerable extension southwards of the Indian Peninsula at a
comparatively recent geological period. Without going into details
as regards this it will suffice to point out the great extent of shoals
and of shallow water lying off Cape Comorin and to the statement
in the ancient Tamil epic " Chilappatikāram " where in the opening
lines of the 8th Chapter reference is made to a terrible irruption
of the sea which devastated a great tract of country to the south of
what is now Cape Comorin The passage states * that the people
of that time (circa second century A D.) had heard from their
fathers that in former days the land had extended further south
and that a mountain called Kumarikkodu and a large tract of country
watered by the river Pahruli had existed south of Cape Kumari,
and that at a time not very long before, in the reign of the Pandyan
King Jayamakirtti *alias* Nilantarutiruvit Pandya, the sea had torn
through the land, destroying the mountain Kumarikkodu and
submerging the whole of the country through which flowed the
river Pahruli.

Lending corroborative weight to this legend are the stories of
similar irruptions of the sea on the south-west coast of Ceylon
recorded in the Buddhist annals of that country. Even now these
stories are current among the Sinhalese of the south, who point to the
outlying rocks known as the Basses, as the remnant of this lost land
which they say was a land of richness abounding in towns and
palaces.

In this connexion too we have to note the significant fact of the
reported presence of large accumulations of oyster shells overlaid
by soil at Muttam, about two miles north-east of Cape Comorin.†

This presence of an old pearl fishery camp within two miles of
the Cape lends further support to the theory of a great extension
southwards of the pearl fishery region and while not conclusive as
evidence add greatly to the physiographical probabilities of such
a hypothesis

At a time when the southern extremity of India extended further
to the south, the pearl banks on both sides of the Gulf of Mannar
would have greater protection from the South-West Monsoon than
they have at present while the extent of suitable ground would be
more extensive. At the present day on the coast of Tinnevelly and

---

* *Vide* V. Kanakasabhai in " The Tamils Eighteen Hundred years ago ", Madras,
1904, p 21.

† *Vide* the statement made to me to this effect in May 1904 by the Jathi Talaivan from
his personal knowledge.

Madura a converse process is in operation, the land slowly gaining upon the sea along the coast line facing the central division of the pearl banks. Two factors are at work—the extension seawards of fringing coral reefs along the coast and the distribution upon the sea bottom of considerable quantities of sand and mud brought down by the rivers Tambraparni, Vaippar, and Vembar There is also a constant movement of sand northwards during the prevalence of the South-West Monsoon, whereby the depth of the water on this pearl bank plateau may possibly be rendered shallower in the central and northern sections

The most recent oscillations of level that appear to have affected the shore line of the Gulf of Mannar and Palk Bay seem to me to be as follows .—

I An early period of depression cutting off Ceylon entirely from India and giving wide and free water communication between what are now the Gulf of Mannar and the Coromandel Coast The land at this period must have been at least 12 to 15 feet lower than now, and probably still more. The whole of the Jaffna peninsula and all of Adam's Bridge, Rāmēswaram and Mannar Islands, and much of the lowlying lands of the present coastal districts of Tinnevelly, Ramnād and Tanjore were submerged. At this period the local varietal divergencies now seen among certain species of marine animals without a free-swimming larval stage, probably did not exist Fossil evidence indicates this depression to have occurred during the Miocene period

II. Next succeeded a period of elevation, when Ceylon was connected with India by a fairly wide belt of land in the Palk Bay area The level of the land was considerably higher than now, at least sufficient to convert the present inland sea of Palk Bay and Strait (now with depths of 7 fathoms over a considerable area) into a shallow marine lagoon, open to the sea probably only at its northern end. It was during this period probably that many members of the Carnatic fauna passed over into Ceylon

III Finally, at a comparatively recent date, a small subsidence occurred, restoring partially the old sea connexion between Palk Bay and the Gulf of Mannar This isolated Ceylon once more and allowed of a new commingling of the marine fauna of the opposite sides of Adam's Bridge. This last oscillation, though comparatively recent and almost within the historic period, was antecedent, I believe, to the formation of the shell deposits in the Tambraparni delta, in Tuticorin harbour and in Pulicat Lake.

V.—NOTES UPON THE DERIVATION AND SIGNIFICATION OF THE NAMES OF THE PRINCIPAL PEARL BANKS OFF THE MADURA COAST.

Number	Names of the Pārs as they appear in the Inspection Reports	Names of the Pārs in the forms now recommended for adoption	Signification.	Derivation and remarks
1	Pāmban Karai Pār.	Pāmban Karai Pār.	Pamban inshore bank, i e., Pamban inner bank	Karai, shore ; pār, rock and by extension of meaning, rocky bank
2	Velangu Pàr	Pāmban Velangu Pār	The bank near Pāmban lying further from shore, i c., Pāmban outer bank	Velangu, further.
3	Musal Thivu Pār.	Musal Tivu Pār	Bank lying near Hare Island.	Musal, hare ; tivu, island.
4	Cholava Karai Par.	Solaka Karai Pār	Bank lying off the south shore (of Rāmēswarim)	Solaka (cholaka) south ; Kurai, shore Cholava probably a misprint for cholaka, which is the fishermen's term for south. "Ten" is the south of the landsman
5	Kilakarai Vallia Malai Velangu Pär	Kilakarai Vellai Malai Velangu Pär.	Outer bank off the white hill at Kilakarai (Kilakarai, literally the East Coast)	Kilai east ; Karai, shore , vellai, white ; malai hill; velangu, further or outer (Kilakarai a port on the east coast).
6	Val'ia Malai Karai Par	Vellai Malai Karai Par.	Inner bank off Vellai Malai.	Karai, inshore, i e , inner in contrast to Velangu, outer Vellai Malai, a white and conspicuous sand dune near Kilakarai.
7	Anna Pär .	Anna Pär …	?	Anna or annam signifies swan.
8	Nallathanni Thivu Pär.	Nalla Tanni Tivu Pär.	Nalla Tanni Tivu bank.	Nalla, fresh , tannir, water ; tivu, island. Nalla Tanni Tivu is an island in the Zamindari of Ramnad where excellent fresh water is obtainable from shallow wells.
9	Upputhanni Thivu Pär	Uppu Tanni Tivu Pär	Uppu Tanni Tivu bank	Uppu, salt An island close to Nalla Tanni Tivu where the wells are brackish.
10	Kumulam Pär	Kumulam Pär	Probably this is "the bank abounding in hemispherical stones."	Kumulam, a blister like swelling is applied to the hemispherical masses of Astræid corals so common on certain sardy oyster banks.
11	Vembar Peria Pär	Vembar Periya Pär.	Large bank lying off Vembar village.	Periya, large

Notes upon the derivation and signification of the names of the Principal Pearl Banks off the Madura Coast—cont.

Number	Names of the Pars as they appear in the Inspection Reports	Names of the Pars in the forms now recommended for adoption	Signification	Derivation and remarks
12	Vaipar Pena Pär	Vaippar Periya Pär	Large bank lying off Vaippar village	....
13	Karai Pär	Karai Pär	Bank lying towards the shore.	
14	Devi Pär	Devi Pär	Probably "the Rani's bank."	Devi, a titular suffix to the names of the Rani of Ramnad and of Madura; primarily the name of a goddess (Siva's wife).
15	Pernandu Par	Pernandu Pär	Fernando's bank	Probably first found by a fisherman of this name
16	Padutha Marikan Pär	Padutta Marikan Pär.	Bank whereon Varikan (a Mussalman) was found lying (dead)	Paduttiru, lying down. Marikan or Marikar appended to Muhammadan names is a relic of an honorofic bestowed by the Portuguese upon prominent Muhammadans
17	Padutha Marikan Pär Thundu	Padutta Marikan Tundu Par	The small bank close to the preceding	Tundu, fragment, i.e., small
18	Tuticorin Coda Par	Tuticorin Kuda Pär	The bank off Tuticorin Bay	Kuda, bay; primarily a cavity or hollow.
19	Cruxian Pär	Cruxian Pär	Bank of the cross	Kroos (cnrusu), a corruption of the Portuguese Cruz, cross
20	Cruxian Thundu Pär	Cruxian Tundu Pär	Small Cruxian bank.	...
21	Vanthivu Arupajam Par	Vantivu Aru pagam Pär	Bank off Vantivu in six fathoms (Vantivu, hard white rock island).	Aru, six. pagam, fathom; van, hard. Vantivu, a small island, north of Tuticorin, on which there is a small fishing beacon; probably so called on account of hard white rock there present
22	Nagraa Pär	Nagara Par	Bank abounding in a fish called nagara	Nagara, the name of a fish
23	Ootti Par	Uti Pär	Snail bank	Oti or Uti, a snail like mollusc. Probably a corrupt form of Uti, the name of a small spiral shell fish (gastropod) that is an active enemy of the pearl oyster during the first four months of life.
24	Ooduruvi Pär	Uduruvi Par	Possibly has reference to a cavernous condition of the rock, enabling fish to pass in and out of the cavities.	Uduruvi, penetrating or passing through.

Notes upon the derivation and signification of the names of the Principal Pearl Banks off the Madura Coast—cont.

Number	Names of the Pārs as they appear in the Inspection Reports.	Names of the Pārs in the forms now recommended for adoption	Signification	Derivation and remarks.
25	Klatti Pār	Kilati Par	Bank abounding in trigger fish.	Kilati, trigger fish (*Balistes mitis* is the most abūnd ant species on this coast).
26	Athuvi Arupajam Par	Attuvai Arupagam Pār.	bank near mouth of the river in six fathoms	Attuvai, mouth of a river (attu, river, vai, mouth aru, six; pagam, fathom).
27	Athombadu Par.	Attonbadu Pār	9 (fathom) bank off the river.	Attu, river; onbadu, nine (ombadu, a corrupt form of onpatu)
28	Pasi Pār	Pasi Pār	Sea-weed bank	Pasi, sea-weed (literally moss).
28a	Patharan Par,	Pattarai Pār.	10½ (fathom) bank	Pattu, ten ; arai, half
29	Kuthadiar Pār.	Kutadiar Iār.	Dancer bank (?,	Kutadian, a dancer
30	Tholayiram Pār.	Tolaviram Pār.	900 ba ks	Tolayiram, 900. A large bank characterized by numerous small patches of rock rising from a sandy bottom
31	Vadda Ombathu Pār.	Vada Onbadu Pār.	Northern bank in nine fathoms.	Vada, north.
31a	Saith Ombathu Pār	Saith Onladu Pār	Southern bank in nine fathoms.	Saith, south (a local term derived I think from Portuguese).
32	Puli Pundu Pār	Puli Pundu Pār.	Bank having a tamarind bush as land mark	Puli, tamarind; pundu, bush.
32a	Canna Puli Pundu Par.	Kanna Puli Pundu Pār	Bank having a tamarind bush as land mark near Kanna Tivu	Do.
33	Alluva Pār .	Alluva Pār	Rotten weed bank (?)	Alluva, rotten (pasi, sea weed being understood)
34	Kanna Thivu Arupajam Pār.	Kanna Tivu Arupagam Pār	Bank off Kanna Tivu in six fathoms	Aru, six ; pagam, fathom
35	Thundu Pār	Tundu Pār ...	Small bank	
36	Nenjurichan Pār	Nenjurichchan Pār	Good for nothing bank.	Nenjurichchan, good for nothing, literally "heart harrower" from Nenju, heart ; Urichchan, flayer or peeler
36a	Par Kudanjan Pār	Par Kundanjan Pār.		
36b	Mela Ombathu Pār	Mela Onbadu Pār	Further or outer 9-fathom bank	Mela, further or distant, outer.
37	Punyacoil Seltan Pār.	Pinnakayal Seltan Pār		Pinnakayal, *Anglice* Pinnacoil, the name of a large Parawa settlement on the Tinnevelly coast
38	Sandamaram Puditha Pār.	Sandamaram Piditta Pār.	Bank where land mark is a tree near a market.	Sanda, a corrupt form of Santar or Chantai, market , maram, tree ; piditta, touching, i.e., indicating as does a land mark.

Notes upon the derivation and signification of the names of the Principal Pearl Banks off the Madura Coast- cont

Number	Names of the Pārs as they appear in the Inspection Reports	Names of the Pārs in the forms now recommended for adoption.	Signification.	Derivation and remarks
39	Ira Thivu Cudamuthu Pār	Irai Tivu Kudamuttu Pār.	Bank in the Pearl Bay near Irai Tivu	Irai, literally finger joint, inch ; kuda, bay · muttu, pearl.
39a	Nadukuda muthu Pār	Nadu Kuda muttu Pār	Bank near the middle of the Pearl Bay	Nadu, middle.
40	Kovilpuditha Puthu Pār.	Kovil Piditta Pattu Pār.	Bank in 10 fathoms having a church as the leading mark on shore	Piditta, touching , kovil, church.
40a	Sanguria Puthu Pār.	Sankuraiya Pattu Pār.	Bank where depth is a foot less than 10 fathoms	San, span , kuraiya, less (short of).
40b	Nilankalla Par.	Nilan Kallu Pār.	Blue stone bank ...	Nilam, blue · kallu, stone.
40c	Sethu curia Pathu Pār.	Sattu Kuraiva Pattu Pār.	Bank a little less than in 10 fathoms	Sattu, a little ; kuraiya, less ; pattu ten
41	Kudamuthi Pār.	Kudamuttu Pār	Pearl Bay bank ..	Kuda, bay ; muttu, pearl
41a	Rajavukku Sippi Sothicha Pār.	Rajavukku Sippi Sotichcha Pār.	Bank searched for oysters for the Raja.	Rajavukku, for or to a Raja ; sotichcha, searched or examined ; sippi, oyster
41b	Saith Kudamuthi Pār	Saith Kuda Muttu Pār	South Bank in the Pearl Bay	· .
42	Naduvu malai Piditha Pār.	Naduvu Malai Piditta Pūr	Bank touching or on the bearing of a peak of the Western Ghats	Naduvu malai, Western Ghats (Naduvu, central ; malai, hill).
42a	Peria Malai Puditha I ār.	Periya Malai Piditta Pār	Bank touching the great hill.	. .
43	Kadian Pār	Kadeiyan Pār	The Lime-burners' bank	The Kadaiyar caste is that of the lime-burners ; it however furnishes a contingent of men who work as divers at the chank and pearl fisheries.
43a	Kanava Pār ..	Kanawa Pūr ..	Bank abounding in cuttle fish	Kanawa, cuttle fish.
43b	Puthu Pār	Pudu Pār .	New bank ..	Pudu, new.
44	Karia Karwal Pār	Karai Kuruwal Pār	In shore black bank	Karuwal. black , probably from the colour of the surface of the rock.
45	Velangu Karwal Par	Velangu Karu wal Pūr	Off shore black bank	. . .
46	Thundu Pār ..	Tundu Pār ..	Small bank ...	. . . .

Notes upon the derivation and signification of the names of the Principal Pearl Banks off the Madura Coast— cont

Number	Names of the Pārs as they appear in the Inspection Reports	Names of the Pārs in the forms now recommended for adoption	Signification	Derivation and remarks.
47	Trichendore Punthotta Par	Trichendur Puntoddam Par	Trichendur flower garden bank.	Puntoddam, flower garden, probably so named on account of some pretty species of sea weed or other marine organism found on this bank
48	Odacarai Pār	Oda Karai Pār	Narrow bank towards the shore.	Odai, narrow.
48a	...	Oda Karai Tundu Par.	Small Odakarai bank	*Vide* 46 and 48
49	Chowdi Pār	Chodi Pār	Ornamental bank	Chodi, to adorn
50	Sandamacoil Puditha Par	Sandamacoil Piditta Par	Bank touching St Mary's Church, i e , where St Mary's Church is the principle bearing	Sandamacoil, St Mary's Church ; Sandama, holy or blessed mother, i.e., St. Mary (Sancta Maria) ; kovil, church
51	Theradi Puditha Pār.	Teradi Piditta Pār	Bank having Trichendur pagoda as land mark	Ter, car, teradi, place where car is kept, i e , pagoda or temple
52	Semman Patti Pār	Semman Path Pār	Red hill bank, red hill is the bearing).	Sem-man, red sand or earth , path, a hill (?)
53	Surukku Ombathu Pār	Surukku Onbadu Par	Bank falling quickly into 9 fathoms, i e., a rapidly shelving bank (this is the case, it is a narrow bank 5½ fathoms deep on the west side, 8 fathoms deep on the east)	Ombathu, nine (a corrupt form of Onbadu) , surukku, quickly.
54	Manapad Peria Pār.	Manappad Peria Par	Large bank off Manappad	Manappad, name of a promontory.
55	Kanawa Paraku Sohi Thundu Pār	Kanawa Parakku Sohi Tundu Pār.	The small bank of the highly coloured flying cuttle fish	Kanawa, cuttlefish ; parakku flying ; sohi, well dressed or adorned
56	Paracherry Pār.	Paracheri Pār	The bank off the Pariah village.	Cheri, village , paraiyan, pariah (drummer)
57	Paracherry Pathoor.	Parachen Pattu Pār	10 fathom bank off the Pariah village (?).	Pathoor appears to be a misrendering for Pattu Pār, the depth being 10 fathoms. *Pathoor* may also signify the "10 villages," but the context does not support this reading
58	Alanthalai Pathoor	Alantalai Pattu Pār.	The 10 fathom bank off Alantalai (?)	
59	Manapad Pathoor.	Manappad Pattu Pār	Manappad 10-fathom bank (?)	

Notes upon the derivation and signification of the names of the Principal Pearl Banks off the Madura Coast—cont

Number	Names of the Pārs as they appear in the Inspection Reports	Names of the Pārs in the forms now recommended for adoption	Signification	Derivation and remarks
60	Keelee Pār	Kili Pār	Bank abounding in parrot-fish.	Kili, the name of a small fish (parrot-fish).
61	Peria Thalai Seman Tharai Par.	Periya Talai Seman Tarai Pār.	Bank lying near the great head (land) of red earth	Talai, head ; seman, red , tarai, earth (Latin, *terra*).
62	Seman Pallei Kathu Par.	Siman Pillai Katha Pai	Possibly this means "the bank guarded by Siman Pillai."	Siman Pillai, a man's name. Katha, guarded by
63	Kodoo Thalai Par.	Kuda Talai Pār	The bank at the head of the bay (?)	... .
64	Ovaree Antho niar Kovil Puditha Pār.	Ovari An honi ar Kovil Piditta Pār.	Bank touching St Anthony's church at Ovari, i e , on the hearing of this church.	Ovari, the name of a village Anthoniarkovil, St Anthony's church.
65	Ovaree Antho- niar Kovil Vallai Velai Pār.	Ovari Anthoni- ar Kovil Vallai Velai Pār.	?	

VI.—NARRATIVE OF THE EXAMINATION OF THE PEARL BANK REGION.

As knowledge of the observations made during the May cruises of the S S. "Margarita" is essential to a full understanding of my recommendations, it will be convenient if I give the record in the character of a narrative altering as little as possible the form in which it stands in my diary

No attempt is made to furnish extensive lists of the animals found. Singlehanded the task is an impossibility, while the limitations of time do not permit the reference of the collections to specialists. I have however expended considerable time and labour upon the identification of those organisms that are characteristic of certain banks—all those that predominate or have special significance are either signalized by name or by description. The present identification is sufficient for the purposes of comparison ; the rarer and smaller animals may well be left for detailed examination at a subsequent period.

The programme of investigation which I mapped out before leaving Ceylon consisted of three main lines—

(a) The examination of the sea-bottom in the pearl-bank region on the Indian side of the Gulf of Mannar, for the purpose of the institution of both physical and biological comparisons between these banks and those with which I am now so intimately acquainted on the Ceylon side.

(b) An inquiry into the present methods in use for the inspection of the banks, and the means, if considered necessary, requisite to render the work of inspection adequately effective

(c) A critical examination of the historical evidence available, inspection and fishery reports in particular, in order to ascertain what localities the past records demonstrate to be more favourably situated than others for bringing pearl oysters to maturity.

The present section deals almost entirely with the first mentioned of these lines of enquiry ; the historical evidence gleaned during the trip is incorporated with other data in another section.

The Ceylon pearl fishery closed by reason of unfavourable fishing weather on April 22nd this year (1904), but in the hope that the Indian coast might enjoy sufficient shelter to permit of useful work being done ere the monsoon came on in full fury, I telegraphed my preparedness to Captain Carlyon, the Port Officer of Tuticorin. Accordingly on the morning of the 26th, the S S "Margarita" arrived at Marichchukadde under the command of Captain Carlyon who also acts as Superintendent of the Indian Pearl Banks. Some hours were occupied in the transfer of baggage and by nightfall we proceeded for Pamban, having in tow three of the Ceylon inspection whale boats

A heavy sea prevailed during the passage, rendering it one of much discomfort On arrival at Pamban the next morning, it was decided to make Tuticorin by the sheltered passage formed by the string of islands that skirt the coast for the greater part of the distance between these two ports.

Anxious to see the important Muhammadan diving community of Lebbais* settled at Kilakarai we put in there on the afternoon of April 27th Accompanied by Captain Carlyon I went ashore

---

* In Ceylon these men are known as " Moormen "

and by the courtesy of the Gomez family we were enabled to interview the headmen of the divers in their house—Xavier Gomez, who had been one of the assistant beach-masters at the Ceylon Fishery, acting as interpreter.

The head diver, M Kirutuneina, who claims to be 70 years old and to have dived from the age of nine, had much to say, but few facts of consequence were elicited, he and the other elders had never known or heard of mature pearl oysters in quantity on any of the Pars between Kilakarai and Pamban The nearest locality they knew of was the vicinity of Nallatanni-tivu, and even there they had never seen a bed of living oysters, the evidence as regards this locality resting entirely upon the abundance of old oyster shells that litter the sand hills of that island

In Kilakarai, signs of prosperity were visible everywhere ; boat-owners were arranging for the building of new craft, their agents gone to Cochin for timber ; diver-fishermen were investing in new nets and goldsmiths were busy with orders for jewellery for the womenfolk. The talk everywhere was of money, of profits past and prospective, individual and collective, and of their determina-tion to send more boats and divers to the next year's Ceylon fishery which they were already exploiting in imagination ! Incidentally the head diver informed us that as the result of careful calculation, the best informed people of the place estimated that this one town had brought away from the Ceylon fishery a sum of fully ten lakhs of rupees —the earnings of the divers, munduks, and boatmen, and the profits of the pearl merchants and boutique keepers—a sum equal to the gross proceeds received by the Ceylon Government from their share of the fishery.

This year the high profits made at the Ceylon fishery were doubly welcome as being unexpected, news having been univers-ally circulated at the end of last year (1903) that, as the result of the November inspection of the banks had proved disappointing, no fishery would take place

The Kilakarai men are the best and most reliable of the local divers who attend the pearl fisheries of the Gulf of Mannar. Their abstemious lives consequent upon fairly faithful observance of the Prophet's laws, predispose to health and regularity of working and while more industrious than the Roman Catholic Parava divers they also make better use of their earnings than do the latter, who, I am assured on all sides—even by their own people—dissipate their fishery gains within a month or six weeks of their return home. Indeed I was told subsequently in Tuticorin, that the great majority of the Paravas will do little or no work till they have got rid of their earnings in drink and in entertainments and are penniless once again

A small settlement of Paravas, dominated as usual by a white-washed Roman Catholic Church, is set within circumscribed limits on the seaward margin of the town of Kilakarai. Few physical differences that cannot be accounted for by the great divergence between their modes of life can be noted between them and their Lebbai neighbours and I incline to the belief that in the Kilakarai Muhammadans (except among the better class families where distinct traces of Arab blood persist) we have the descendants of Tamil fisher converts to Islam, just as the Paravas have become Roman Catholics. Indeed I cannot help thinking that the Paravas and Kilakarai Lebbais are identical in origin, but in the absence of anthropometric measurements the point cannot be settled definitely.

Leaving Kilakarai the next morning we proceeded direct to Tuticorin, landing there on the afternoon of April 28th.

The ensuing three days were spent in completing the necessary preparations for work at sea, getting coal and water aboard the steamer, and, on my part, in interviewing every resident in any way likely to have shrewd opinions based upon local intimacy with the pearl bank region and in comparing and abstracting the information contained in the fishery and inspection records. Unfortunately the latter are all of comparatively recent date, none going back to the period of the Dutch occupation—a *lacuna* which I was subsequently able to fill in great part by the collation and collection of references which occur incidentally in various and diverse publications.

The Government records give the names of over 60 pārs which are reckoned as potential oyster banks. Reference to charts I and II in the appendix, shows that the majority are massed offshore between Tuticorin and Trichendur in from 6 to 10 fathoms of water. This region includes practically all the banks that have yielded fisheries during the present century and accordingly it was decided to make our first cruise over the area thus indicated.

TOLAYIRAM PĀR —Accordingly on May 2nd we left Tuticorin at 6-30 a.m. and proceeded to the south end of the Tolayiram Pār,

9

about 8 miles to the east of Hare island There we commenced the examination, the day's work extending over the southern half of the bank, with traverses extending some distance beyond the charted margin This bank, which yielded fisheries in 1784, 1787, 1807, 1810, 1822, 1830, 1889 and 1890, has generally been considered one of the most favourable for rearing oysters to maturity, and to be fully the equal of any other bank in respect to the number of spat falls reported upon it during the past half century.

It possesses by far the largest area of any productive Indian bank, its charted outline being 7 miles long with a width varying from one mile to two miles The depth varies from 8 to 11 fathoms. Our examination showed the bank to consist of a somewhat uneven, but not rugged, rocky framework rendered level by the accumulation of sand in the depressions Here and there the rock shows bare save for a thin veil of sand, but the greater part is covered by sand varying from 1 and 2 inches to 6 inches and a foot in depth.

The sandy bottom appears to the divers as broken up by a multitude of rocky outcrops usually of limited extent and from this circumstance we may infer the origin and propriety of the name "Tolayiram," literally " nine hundred."

The surface of the bank shows considerable local diversity—both physical and faunistic In some places a rocky surface sprinkled lightly with sand bears loose blocks of calcrete (recently formed rock) of varying size; elsewhere fragments of Madrepore coral branches, corroded and water-worn, lie loose, here sparsely scattered, there abundant. In other places deep sand, bare of any life, largely preponderates Variation in every proportion is represented

The sand is altogether different from that on the Ceylon side Instead of being clean large-grained quartz grit, as there, the sand of the Tolayiram Par is fine in grain, the angles well rounded; chemically it is composed principally of calcium carbonate—comminuted shell fragments in the main.

In colour it is yellowish brown and there is always associated with it a certain, though variable, amount of mud particles, which rise with every movement upon the bottom—the scramble of the divers, the under-tow of strong currents

The majority of the diving descents made upon the bank proper showed the greater part of the area examined to be thickly covered

with young oysters from one month to six months old, those of
three to four months of age preponderating The sizes varied
from $10 \times 10\frac{1}{2} \times 3\frac{1}{2}$ millimetres to $24 \times 22 \times 8\frac{1}{2}$ millimetres The
weight of 100 individuals of average size was 99.65 grammes

The general facies of the bank approximates closely to that of
the Ceylon Periya Pār—a bank noted for the frequence of spat
falls upon it, both are of great extent and of diversified character
and both lie all but out of sight of land towards the edge of
soundings Great quantities of young oysters were found on the
Ceylon bank named in March last, practically of the same age as
those on the Tolayiram Pāi, but on the whole the abundance was
distinctly less on the latter bank, while the sand leaves less extent
of rock exposed

Considerable destruction of the young oysters was apparent,
and large numbers of empty shells were found Of the latter a
small proportion, 1 in 14, bore evidence in the presence of circularly
bored holes, to destruction by small carnivorous gastropod molluscs
(belonging to the genera *Purpura*, *Nassa* and *Sistrum*) termed *Uri*
by the divers The great majority, however, furnished no
indication to show by what agency death had been caused Consideration of what the chief harmful factor is and how it acts will
be dealt with when we deal with the conclusions.

Characteristic organisms are few in number, sponges predominate, the black crests of *Spongionella nigra* being frequent
wherever the sand thins away. The pink *Petrosia testudinaria* is
also common, its truncate massive pile increasing the resemblance
to a miniature volcano by possessing a crater-shaped excavation
upon the summit. Other massive but less conspicuous species are
equally abundant, and in some cases, I found the rapid growth of
these sponges entailing the destruction of many young oysters,
enwrapping and smothering them, as evidenced by the empty
shells embedded in the sponge mass *Axinella donnani* is occasionally met with

Corals were scarce Occasionally locally-isolated colonies,
usually small in size, were found, nearly all being *Porites*,
Meandrina, and Astræids

Still less common were Alcyonarians, represented by *Sarcophyton*
Vermetus was common on the fragments of calcrete, and the
lovely star-fish, *Pentaceros lincki,* a known enemy of the pearl
oyster, was present in considerable numbers *Linckia laevigata*

was also taken, with *Antedom palmata* in crevices of the exposed rock. Little or no alga was present.

In our traverses zig-zag across the bank we several times passed beyond the margin of the par on the westward side, finding there bare and barren sand with an occasional chank (*Turbinella pirum*).

After completing the day's work we anchored on the south end of the par in nine fathoms, and within an hour the crew had caught 16 *Kilati* (Trigger-fishes, *Balistes mitis*). Several were examined and in the stomachs of all were found fragments of several kinds of shells, those of young pearl oysters predominating in many.

The *Parmandadai* or "Rock-pilot" who is taken out by the Inspector to help in locating the banks informed me that this bank has always been noted for the great abundance of Kilati ; one ballam often brings back a catch of 100 fish from this neighbourhood.

During the day the current set strongly from the north, the steamer drifting rapidly when not steaming. The temperature of the sea at 7 a.m was 87° F. and at 5 p.m. 88° F. The specific gravity was 1,022 80 at 7 a.m.

The following morning, May 3rd, on heaving the anchor up, 18 pearl oysters approximately 6 weeks old, all being of the same size, were found attached to the chain near the anchor end, each by several strands of byssus. All were attached to that part of the chain which would occasionally rest upon the ground—the last two fathoms. It is noteworthy that none of the older sizes were found on the chain, although varying ages were found on this spot.

UTI PĀR GROUP—This region, comprising Uti, Nagara, Uduruvi, Kilati, Atuvaiarpagam and Pattarai Pārs, was next examined first by means of diving traverses on May 4th and on the ensuing day by the method of circle inspection as used on the Ceylon side and which is described in detail *infra* on page 156

The pars in this group are small and with advantage may be considered as one, under the above title—Uti Pār Group. The depth is less than the mean on the Tolayiram, averaging here from 7 to 8½ fathoms.

In all, nearly 350 dives were made in the course of an examination by circle inspection. The centre of the circle was fixed at a point which Captain Cailyon believed from the bearings to lie just

on the east margin of the Uti pār, but that bank the pār
mandadai said was really south-east. The results proved the
latter to be right and the Inspector to be wrong, a false position
for the latter to occupy and one that is due entirely to the fact
that *no shore marks are indicated on the chart in use* No Inspector
can be expected to do good work under present conditions

The matter presents no difficulties, the landmarks, which
consist of Haie Island Lighthouse, the factory chimneys at
Tuticorin and the church on Vantivu, are clearly distinguishable
from these banks, and I should think the survey office could, from
the materials already available plot the position of the several
objects with exactitude and without further survey.

The whole of the Uti pār was gone over together with the
southern portion of the Nagara pār, the result showing that but
a few odd living oysters remain, aged from $2\frac{1}{2}$ to 3 years old and
all more or less over-grown with sponges and other growths In
several cases the largest and most frequent of these crusting
sponges, the brick-red *Clathria indica*, completely enveloped the
oysters, occupying the whole surface of both valves and rising in
numerous bold upgrowths to a height of two and three inches.

The pārs constituting the Uti group have absolute identity in
fauna and in physical characteristics. The rock of each pār is
fairly continuous in its outcrop, with much less sand sprinkled over
it than in the case of the Tolayiram Pār To some extent as a
consequence of this the fauna is richer in the number of species,
in the number of individuals, and in luxuriance of growth

Sponges are especially abundant Among the most characteristic are the black-crested *Spongionella nigra*, one specimen of
which was partly mantled with a thin crust of crimson-lake
Botrylloid; purple-red *Siphonochalina communis* (Carter) bearing
frequently a like-tinted *Antedon*, clinging to its tubular branches;
the massive *Suberites inconstans* and the oyster-crusting *Clathria
indica*

Several of these sponges, notably *Siphonochalina* and *Suberites*,
furnish free quarters to quite a host of diverse lodgers—chief
among which are a colourless *Alpheus*, a scarlet *Porcellana*, a small
Gebia and a long-armed spiny Ophiuroid (*Ophiactis savignii*). The
last named chiefly affects the large canals of the *Suberites*, more
rarely being found within *Siphonochalina*. *Gebia* burrows in the
smaller canals of the *Suberites*, while the *Alpheus* and *Porcellana*

favour *Siphonochalina* much more than they do *Suberites*, probably because their superior activity enjoys greater freedom in the larger and less tortuous cavities of the former sponge.

Corals and Gorgonoids are scarce, the handsome *Juncella juncea* being the only conspicuous representative found

Specially characteristic are enormous numbers of the branched parchment-like tubes of a fine Eunicid (*E tubifex*, Crossland). The empty tubes were made known to science years ago by Professor McIntosh from material received from Mr E Thurston, Superintendent of the Madras Museum, but it is only in the present year (1904) that the animal has been described and named. Quite a host of smaller creatures settle upon the surface of the tubes— hydroid zoophytes, polyzoa, and compound ascidians, together with an occasional *Lepas* of a species not yet identified The last named is of interest in that in colour and outline its appearance approximates so closely to a branch of the Eunicid tube that this may be regarded as a striking case of mimicry or protective adaptation to environment.

Colonial masses of the delicate calcareous tubes of *Filigrana* were met with and numerous species of the usual errant worms.

Of Echinoderms, Ophiuroids, *Antedon* spp., and *P linki* were abundant. Many of the Antedons appeared, as already noted, to be commensal (?) with sponges and with Gorgonoids, while commensal in turn with Antedon I found Decapods and Ophiuroids— the former consisting of a small striped crab and a striped Galatheid, the latter of a small, short-armed black Ophiuroid upon an Antedon of the same hue.

Small Cephalopods (*Polypus* spp) were numerous , polyzoa and tunicates were universal

Occasional individuals of a large *Pinna* sp. were found lying prone on the rock and much enveloped with sponge and tunicate growth, barnacles and the like ; some bore pearl oysters of about one year old

To the west of the Uti group of pārs a large chank-bed is marked Here we found rock to be practically absent with a corresponding absence of the faunistic elements noted above. In their place were quantities of chanks (*Turbinella pirum*) and of Pinna shells. The former were mostly small as is to be expected, this being a recognized chank-bed and within easy reach of Tuticorin. Many of the *Pinna* were dead shells , those that were alive were, as is usual on sandy ground, embedded deeply in the sand. Both dead and living bore quantities of large barnacles (*Balanus* sp.)

The rock is of somewhat variable composition, varying from a compact brown limestone, similar to that of the Tolayiram Pār, to rock of a distinctly quartzose nature, the angular quartz grains were embedded in a brown calcareous matrix—a quartzose limestone. The sand of the chank-bed is similar to the component material of the pār, but containing an appreciably greater amount of recognizable shell fragments.

After completion of the inspection of the Uti Pār region, we proceeded south on the afternoon of May 4th in the face of a stiff breeze and heavy sea, anchoring at 5-30 p.m. off Pinnakayal in the shelter afforded by the reef off Kayalpattanam point.

Pinnakayal is one of the headquarters of the Parava caste and a noted Roman Catholic centre. Here St. Francis Xavier laboured with great effect and of the four churches which render the town conspicuous from the sea one is connected by legend with this great missionary's ministrations. It is of Pinnakayal that de Faria y Souza records that (*circa* A D 1560) the Viceroy of India "sailed to the Island Mannar, where he built a fort and translated thither the inhabitants of Punicale to redeem them from the tyranny of the Nayque, who would fleece them there—Emmanuel Rodrigues Coutinho was left to command there and with him some Franciscans and Jesuits, all satisfied with the equal distribution the Viceroy made of all things." *

A Casuarina-tope is a conspicuous feature of the landscape about two miles south of the town and were its position fixed with accuracy upon the chart it would form a useful and much-needed landmark during the inspection of the pearl banks.

The Jathi Talaivan has informed me that an old pearl fishery camp at one time was situated just south of the trees of the tope as evidenced by this place being now called Silavaturai kadu (jungle), the site having now reverted to jungle.

KARUWAL GROUP —The next morning an oily calm prevailed with current running from the north. We steamed south-east with the intention of examining the group of banks lying off Trichendur and of which the Velangu and Karai Karuwal Pars are the

---

* " History of the Discovery and Conquest of India " translated by J. Stevens, 1695, and quoted in the Ceylon Monthly Literary Register, Volume III, N S , p 199

central and among the most important, having given fisheries more frequently during the past century, than any other section of the pearl banks, with the single exception of the Tolayiram Pār. *

At 8 a m the four inspection boats were cast off to the south-west of the Velangu Karuwal Pār at a point due east of Trichendur Pagoda. The boats were ranged in line abreast, a quarter of a mile separating the individual boats and the coxswains were instructed to follow the steamer taking dives at regular and frequent intervals and preserving their respective distances apart. We then steamed three miles north by west and anchored on the west side of the Naduvu Malai Piditta Pār in $9\frac{1}{2}$ fathoms, the current still running strong from the north and the wind remaining southerly

When the boats arrived, it was found that only the two on the west of the line had crossed over the pārs, and others being too much to the east and traversing ground which was almost entirely bare sand   The results obtained showed the rock and sand to be of the same characters as the bottom on the Uti Pār region ; in some places upon rocky ground a considerable amount of Orbito-lites sand was found and in other places the sand was coarse enough to be considered a gravel. The fauna was in its main characteristic features similar to that of the Uti Pār—sponges were abundant and of similar species and, in addition, several specimens of the spherical crimson *Arinella tubulata* were obtained, containing the usual quota of commensals--Oligochæte worms and Gephyreans

Of corals we found *Favia* sp. forming rounded masses 5 to 8 inches in diameter. The tubes of *Eunice tubifex* were again common together with *Pentaceros linki*, *Antedon*, Ophiuroids and many Polyzoa, the most conspicuous of the lastnamed being dense hydroid-like colonies of *Scrupocellaria* sp. over 3 inches in height. Little seaweed was found but *Padina commersoni* was sometimes fairly common together with some bunches of *Codium tomentosum*. On Naduvu Malai Pār small nullipore balls (*Lithothamnion*) were locally abundant on certain of the sandy stretches. Pinna and live coral were absent from the ground examined this day.

The sandy ground to the west of the pārs yielded numerous chanks and many valves of sand-loving Lamellibianchs (*Mactra*,

---

* The Karuwal Pārs were fished in 1805, 1815 and 1862

etc.), the sand itself being of the usual brown calcareous nature, of fine grain and with comparatively little quartz

A few living oysters were found on the Naduvu Malai Pār aged from 2½ to 3 years together with many dead shells of about the same age, largely on sandy bottom As in the Uti Pār region the majority of the shells were enveloped in a covering of sponge (*Clathra indica*)

The landmarks for the Karuwal Pār region are excellent To the south Manappad lighthouse is conspicuous, due east is the lofty pile of Trichendur Pagoda while to the north-west is the white mosque near Kayalpattanam village and a Casuarina tope to the south of Pinnakayal, to say nothing of a white gabled Roman Catholic Chapel at or adjacent to Kayalpattanam point Unfortunately the three latter are not marked upon the chart and as the Inspector was uncertain as to whether the tope is on an island or on the mainland and as to the exact relative positions of the chapel and the point named, the difficulties which I experienced in localizing the boundaries and positions of the pars were great and distracting

From what I have seen already and from the silent evidence afforded by the charts on which the marks are either not placed or are indicated vaguely and without precision I am convinced that the work of inspection for years past has been carried out without that scrupulous exactitude necessary to obtain satisfactory and reliable results

INNER KUDAMUTIU PĀR GROUP.—This group, consisting of the Saith Kudamuttu, Kudamuttu, Rajavukku Sippi Sotichcha, Sandamaram Piditta, Pinnakayal Seltan and a few other small pārs, was examined on May 6th, the steamer accompanied by the four inspection boats, two on either side, proceeding slowly from one end to the other, the divers descending to the bottom at regular intervals

The ground upon the pārs appeared less favourable to the maturing of pearl oysters than that of the Karuwal group Competing organisms were in greater numbers and more luxuriant in growth , the banks were typically "dirty," using the term in the oysterman's sense of being pre-occupied by organisms, sponges especially, which give no opportunity to the well-being of oyster spat settling thereon

A few old oysters apparently 2 to 2½ years old were found, less than half a dozen in number, together with some of a younger

10

generation, 9 to 10 months old. The majority of both ages were, as usual, densely covered with sponge growth (*Clathria indica*) Dead shells of the younger generation were in quantity in some places Death in many cases had been recent and the majority of these showed distinct signs of having been bitten, pieces having been snipped out of the ventral margin suggestive of damage by oyster-eating fishes, which were found notably numerous here, seve ι *Kilati* (*Balistes mitis*) being caught by the No. 4 boat which traversed the greatest extent of rocky bottom, while in the evening after anchoring, several large Vellamīn (*Lethrinus* sp.), noted devourers of shellfish, were taken together with more Kilati.

Although we found no pearl oysters in quantity, the work of the boats showed that the ground to the east of this group and between it and the Sankuraiya Pattu Pār group is excellent chank ground and should be marked as a chank-bed on the chart. It will probably be found to extend also some distance northwards The chank-bed sand was excellent of its kind—fine grained and very dark in colour, due to the presence of mud and organic particles and so forming an excellent feeding ground to the annelids which constitute the favourite food of the chank.

Seaweeds were common on these pārs, principally *Padina* and the lamellar olive-brown fucoid so characteristic of the Ceylon Periya Pār

The characteristic fauna consists of—

Spongionella nigra, Suberites inconstans, Clathria indica, Axinella tubulata and *Siphonochalina* (with the usual commensals) as the most conspicuous and numerous sponges; a coarse form of the decalcifying sponge *Cliona*, making burrows of large size, is also conspicuous in the blocks of dead corals occasionally met with.

Eunice tubifex, Trophonia and many small Polynoids and Serpulids, with Gephyreans and Nemertines

Dromia sp., *Alpheus* sp., *Gebia* sp., *Squilla* sp. and numerous other small Decapods

Pentaceros lincki was present in quantity with an occasional *Linckia miliaris*, and numerous *Antedon* spp and Ophiuroids

No *Pinna* was taken either on the sand or the rock

Ascidians were scanty in number.

The pār is flat-surfaced and in places discontinuous, varying from quartzose limestone to compact and extremely hard, brown, and purely calcareous rock. The loose fragments numerous in certain localities are either of the latter character or are masses of

dead coral much tunnelled by boring molluscs and sponges
From the absence of live coral on these pars, I am of opinion that
in common with the loose broken coral branches ("chullai or
challai" as the Tamil divers term the latter) these coral fragments
are derived from inshore reefs, of which a long one stretches northwards, parallel with the coast, from Tiruchendūr to Pinnakayal

May 7th was spent in making traverses over the two groups of
pārs lying south-west of the Tolayiram Pār—an inner, which we
may term the Puli Pundu group, and an outer, or Nenjurichchan
group. The ground between and around was also examined.

PULI PUNDU GROUP —This collection of small pārs, comprising
the Kanna Puli Pundu Pār, Puli Pundu Pār, Saith Onbadu Pār and
Vada Onbadu Pār lie close together and agree in all essential
characteristics, in the depth of water, which ranges from $7\frac{1}{2}$ to $8\frac{1}{2}$
fathoms, in the identity shown by the organisms found there
and in the physical nature of the rock forming the bottom
For the practical purposes of inspection and fishery they may be
considered as a single unit   They agree exactly in all particulars
with the pārs forming the Uti Pār group, except that at the present
examination no pearl oysters were found

NENJURICHCHAN GROUP.—Very different are the banks which
I propose to unite under the term Nenjurichchan group   The
constituent banks are the Pār Kundanjan, Nenjurichchan and Mela
Onbadu Pārs with a depth ranging from $7\frac{3}{4}$ to $8\frac{1}{4}$ fathoms
These pārs, although in almost the same depth of water as those of
the Puli Pundu group, bear a fauna more characteristic of deeper
water conditions, *Gorgonia miniacea, Suberogorgia suberosa,* with
numerous examples of *Juncella juncea* being characteristic As
usual, on these banks massive sponges are numerous, mostly dark
coloured, and the rock, instead of being covered with ordinary
sand, is sprinkled freely with large foraminifera (*Orbitolites* and
Heterostegina) ; the rock is flat in surface and both in appearance
and in fauna bears much resemblance to the seaward side of the
Ceylon Muttuvaratu Pār

Both to the west, the east, and the south-east of this group
we have a great extent of sandy ground extremely rich in life,
characterized by the presence of varied forms of Alcyonarians.
Conspicuous among these last are the rosy tinted aborescent *Soleno
caulon* (see page 87), a grey and drab *Pennatula* sp and a slender
Virgularia (*V juncea*), the first named anchored by a branching root-
like base, the two latter by a long and deeply embedded axis.

Equally characteristic is a stout lamellar and fan-shaped dark green Alga which passes below into a bulbous base embedded deeply in the sand, the projecting fan-shaped portion measures in many cases as much as 5 inches across; the bulbous base from 6 to 8 inches in length, stiffened by a large admixture of sand entangled among the ramifying filaments of the Alga

That peculiar Ascidian, *Rhabdocynthia pallida*, is another characteristic organism

Small crustaceans, molluscs and burrowing worms are also noteworthy, with a varying number of chanks.

After examining this region we proceeded N N.W from Par Kundanjan Par which brought us to the region immediately southwest of the Tolayiram Par Here we found considerable quantities of small oysters lying in clusters upon sand; the age appeared to be from three to four months and, besides the living, a considerable quantity of dead shells was found A few of the latter showed signs of having been bored by carnivorous Gastropods and others were broken, possibly by the bites of some fish This, however, is somewhat doubtful and the majority showed no apparent signs of what the cause of death had been

The nuclei of the clusters were in the main the spicular Ascidian *Rhabdocynthia pallida*; in a few cases only was it a shell, a fragment of par or of dead coral In many a nucleus was absent, the little oysters clinging to the shells of one another Everywhere there was a marked scarcity of "cultch" (shells and rock fragments)

The sand was clean and in some places contained a larger proportion of quartz grains than at any other place hitherto examined *Pennatula* sp. and *Virgularia juncea* were fairly common as was also the flat echinoid *Clypeaster humilis* A few *Pentaceros lincki* and *Fungia dentata* were also met with

From this position we went west, traversing the area indicated on the chart as a large chank bed. Here the sand became fine, dark in colour and slightly muddy. Zigzagging over this 16 dives were made, giving uniform results regarding the character of the sand and the organisms characterizing it, numerous small chanks, small *Pinna* sp. and rooted fan-algæ

Our supply of water was by this time almost exhausted; so after completing the examination of this chank bed, the "Margarita" was headed for Tuticorin where we arrived the same afternoon.

The two next days I spent ashore in a further examination of records and in gathering local opinions, which I hold should never be treated with indifference. A mean has to be steered between the extremes of credulous faith and scornful contempt, and if the sifting be judicious the stories and opinions of fishermen and divers may often furnish useful hints of considerable importance in drawing deductions and in furnishing the necessary clue to elucidate some difficulty or apparent contradiction

A study of the significance and origin of place names may also furnish considerable assistance, and the Jathi Talaivaimore, whose title is frequently rendered as Jathi Talaivan, was fortunately able and willing to facilitate my task.

Under Captain Carlyon's kind guidance I was enabled also to examine the godown in which the chanks collected by divers on Government account are stored pending the periodical auction sale.

This store is situated about a mile north of the town at a spot conveniently near the shore, on land that once was a salt marsh.

The great majority of the shells were of medium size; quite a large number bore a cluster of young oysters, two to six months old, upon the upper whorls of the shell. In one case 14 young oysters had been so carried – a fact bespeaking both the abundance of oyster spat during the last six months and the poverty of resting places for the attachment of the spat at the end of the free-swimming stage.

Chanks of a size too small to be paid for, formed a heap of quite respectable dimensions.

The "Margarita" left Tuticorin on her second cruise at 7 a.m. on May 10th The weather showed signs of impending change and on that account I was extremely anxious to ascertain at once the condition of the northern pars in order to compare with that of the southern groups with which I had now obtained a fairly satisfactory acquaintance I decided therefore to devote this cruise to an examination of the banks between Kilakarai and Tuticorin.

CRUXIAN GROUP —Of these banks the three southernmost, the Vantivu Arupagam Par, the Cruxian Par and the Cruxian Tundu Par, may conveniently be grouped together as the Cruxian group.

They have long been classed among the banks from which a fishery may from time to time be expected* and accordingly, although they cover a comparatively small area, I made a specially

---

* The Cruxian group gave satisfactory results at the fishery of 1861.

exhaustive examination first by a traverse from end to end by the four inspection boats strung out in line at quarter mile intervals as usual and upon the completion of this, by causing them to make four circles round the ship when at anchor upon the centre of the most important of the group—the Cruxian Par itself.

At 8 a.m. a half knot current was running from the north, with wind from N.W.; temperature of water 89° F.; specific gravity 1,022 80

Besides the work done by the divers in the boats who made in all 250 descents, a large number of check dives were made from the steamer. These, when compared with results obtained from the boats, showed the three pars to have a distinct and characteristic facies of their own, and pointed to the practical advisability of uniting the three under one head both on account of faunistic and of physical identity.

The rocky ground was flat-surfaced and largely continuous ; the sandy stretches found on and between the pars were never deep, scarcely ever exceeding a depth of 3 inches, and in consequence, individuals of the large species of Pinna, which is here characteristically abundant, are exposed for fully five-sixths of their length, only the apex being embedded in the sand. Quite a large proportion were dead shells.

Crowds of large Barnacles (*Balanus* sp.) and Zoophytes occupy the outer surfaces of the valves as well of the living as of the dead, the cavities of those barnacles that are dead harbouring great numbers of small crustaceans and worms.

A favourable feature of these pars is the absence of an excessive amount of sponges Such as there were of the larger forms consisted principally of the massive *Suberites inconstans* and the cavernous *Siphonochalina communis*. In the former, besides the usual species of *Gebia*, were several of an interesting heteronereid form of *Nereis*. The tall, branched tubes of *Eunice tubifex* were met with wherever rock appeared on the surface

A characteristic organism is a *Botrylloides* sp. which forms grey gristly-looking rounded masses of 3 to 4 inches in diameter Algæ were scant in quantity.

The depth of the water over these pars shows great regularity, ranging within the limits of half a tathom, 6 to 6½ fathoms.

The only signs of pearl oysters consisted of occasional dead valves, old and much corroded and of an age which I would fix approximately at 2½ years. They had the appearance of having been dead several years.

A few chanks were found on the sand outside the eastern margin of the pars, where the sand is fine and largely calcareous in composition

From 9–30 to 11–15 a.m. an oily calm prevailed, the surface of the sea covered as far as the eye could see with a brown scum composed of *Trichodesmium erythræum* Its appearance and to some extent its effects were such as a thin film of oil produces when spread upon water

We remained at anchor till the next morning near the middle of the Cruxian Par. A strong southerly swell prevailed the whole night and as a land wind blew with some force from the N W. the ship rolled heavily making sleep practically impossible

At 7 a m. on May 11th we proceeded northwards after casting off the four inspection boats, which were ordered to examine the ground lying between the Cruxian Par and the most southerly of the next group to be examined —the Marikan Par group—and then to pass northwards over the whole extent of the latter pars

The steamer led the way dropping a mark buoy on what was believed to be the south side of the Padutta Marikan Tundu Par, as a guide to the following boats. We then headed north and anchored on the Devi Par

The boats reported continuous sand after passing the northern edge of the Cruxian Par, both between the two groups and along the line which should have led them over the Marikan Par group of banks—an unsatisfactory result due either to the plotting of the cross-bearings being incorrect or to the pars being incorrectly placed upon the chart I incline to believe that the former explanation is the true one, seeing that the Inspector, although the shore-marks are distinct and clearly visible, has not the position of these marks indicated upon his chart Rough and inexact work is the inevitable consequence.

VAIPPAR KARAI PAR.—The next day, May 11th, after ascertaining that the ground in 6½ fathoms, two miles east of outer Challai Island, consisted of fine sand mixed with a considerable amount of mud, we proceeded towards the south-east and anchored upon Vaippar Karai Par in 6¾ fathoms in order that I might make a descent in the diving dress.

When I alighted upon the bottom I found the water so turbid with suspended mud particles that it was impossible to ascertain what lay underfoot save by crawling on hands and knees, and even then I had to bring the helmet window within two inches of the ground. By so doing I found the bottom to be composed of fine calcareous sand commingled with a considerable proportion of muddy sediment. The surface was littered with numerous dead pearl oyster shells both entire and fragmentary, and was underlaid at a depth of from 1 to 2 inches by flat-surfaced rock. On the least disturbance the mud constituents rose in dense clouds further obscuring vision. No recognizable reason for the death of pearl oysters could be traced; their average age was approximately $1\frac{1}{2}$ year. Only a single individual was found alive; it appeared rather older than those that were dead, with the valves covered with a dense coating of tunicates, sponges and polyzoa; it had a distinctly stunted appearance. I also came across a considerable number of large *Pinna* lying prone on the surface, covered as in the case of those found the day previous, with quantities of large barnacles (*Balanus* sp.) together with a mantling of various species of Leptoclinids and several species of zoophytes. Several massive corals of the kinds usually associated with pearl banks were noticed.

This bank, which is not marked upon the inspection chart, dated 28th November 1892, by the bearings should lie between the Vaippar Periya Pār and Pernandu Pār as shown on that chart, but concerning this position there appears to be some doubt as the pār-mandadai (native pilot) who accompanied us held that in reality it lies to the north-west of the Devi Pār.

From the character of the bottom such view is not unlikely to prove correct, for however carefully angular bearings be taken, the imperfections of the present working chart hinder them from being plotted thereon with accuracy.*

The Vaippar Karai Pār is an exceptionally dirty and muddy bank, wholly different from the Cruxian Pār and associated banks. From what I saw and also from the history of the banks it appears, however, to be rather favourably situated for the deposit of spat

---

* The compass bearings were difficult to fix owing to the heavy and continuous rolling of the ship. As near as they could be made out they were

Hare Island Lighthouse, S. 57° W.
Church Gable, Tuticorin, S. 77° W.
North end, Challai Island, N. 17° W.

and appears to be fitted to bear them till they reach from 1 to $1\frac{1}{2}$ year old, after which age they rapidly die off. Such a bank would be a fitting one to utilize as a source from which to obtain young oysters were oyster transplantation ever to be attempted

In the afternoon a heavy squall came on suddenly from the west-south-west, raising a heavy sea. The wind remained in the same quarter all night and the next morning the sea was so rough and the ship rolling so considerably as to render it impossible for me to make further diving descents. A native diver who went down reported the water too thick to permit him to see anything. After waiting a while in the hope that the weather would moderate, we proceeded to

NALLA TANNI TIVU, anchoring off the west shore. Here we landed soon after to verify, if possible, the presence of the oyster shells reported to us by the divers at Kilakarai. Much of the island consists of sand dunes overlaying a coral formation in which can be traced specimens of corals of the same species as those now living upon the adjoining and encircling reef. The island is farmed under the zamindari of Ramnad and several flourishing plantations of casuarina, coconut and palmyra were here found We searched the sand dunes on the western side carefully and in several places we certified the presence of large quantities of pearl oyster valves, both entire and fragmentary; the nacre was undimmed in the case of many and even the mottled prismatic outer coating was intact in some, showing even the characteristic radiating purplish-brown bands distinctly. These shells certainly represent the remains of a fishery camp held here, when, it is impossible to say from anything in the outward appearance of the shells, as they might remain unchanged and uncorroded for an indefinite period when covered with sand in the comparatively dry climate of this locality.

The age of these shells when fished was, judging from the breadth of the hinge groove, not less than $4\frac{1}{2}$ years, possibly five years

There is no record of any fishery camp having been held here under British or Dutch control, and it has been suggested as probable that these shells on Nalla Tanni Tivu represent a fishery held by one of the Ramnad Rajas. This theory may, however, be dismissed at once as untenable, for we have no evidence that these local potentates ever claimed the right to fish pearl oysters

11

in this district, though the zamindari does maintain the right to rent out the local chank fishery at the present day ; we have direct and overwhelming evidence that both the Portuguese and the Dutch, over a period of nearly three centuries, exercised the sole sovereignty over the whole of the pearl fisheries on both sides of the Gulf of Mannar.

The more probable explanation is that these shells represent the remains of a Portuguese pearl fishery camp located here *circa* 1560-70, during the period when the Portuguese, at war with the Nayak, blockaded the Madura Coast and removed the Paravas from Tuticorin and Pinnakayal to settlements in the islands at the head of the Gulf of Mannar. This particular island of Nalla Tanni Tivu would be the natural location of a camp to serve a fishery off the Indian Coast under such circumstances, as it is the nearest one affording a satisfactory and sufficient fresh-water supply.

NALLA TANNI TIVU AND UPPU TANNI TIVU PĀRS.—Early on the morning of May 13th we left our anchorage off Nalla Tanni Tivu and steamed south 2½ miles to the south-west end of Uppu Tanni Tivu Pār, where we cast off the four inspection boats with instructions to row E.N.E. over this bank. Several dives from the ship in five fathoms at this place showed the bottom to be flat rock with a considerable amount of sponge growing upon it. From this place we steamed to the eastern edge of Nalla Tanni Tivu Pār and there awaited the arrival of the inspection boats

The results showed that the interval of sandy bottom between these pārs as shown in the inspection chart is largely absent ; that the extent of hard bottom is more extensive than is charted, and that special attention should be given to this region at inspections, especially in view of the remains of old shells on the neighbouring island of Nalla Tanni Tivu. The depth of water on Uppu Tanni Tivu Pār varies from 4¼ to 6 fathoms, while that on Nalla Tanni Tivu Pār ranges from 5½ to 7 fathoms, depths rather greater than those recorded on the inspection chart

The rock on both pārs varies from an almost pure limestone to a calcareous sandstone, in the former case brownish yellow in colouring and ringing like iron under a blow ; this I consider the bed rock of the plateau and not a recent calcrete. It is apparently identical with the hard limestone of Manappad headland.

Few specimens were obtained as the water was too clouded with mud to permit of objects being seen upon the bottom The divers complained bitterly of the discomfort of these conditions under which they had to examine the bottom by touch alone.

This excessive turbidity is in itself quite sufficient to entail upon pearl oysters starvation, weakness and eventual death, especially in the case of young and immature ones. A similar condition entailing fatal results I noticed among the younger oysters which I kept in aquarium tanks at Galle during the south-west monsoon when discoloured turbid water is the prevailing condition in Galle harbour.

My coxswains, who have been connected with Ceylon inspections for a very long period, 12 to 19 years, state that while the water becomes discoloured on the Ceylon side after a continuance of heavy weather the extent of turbidity is slight compared with what they have seen on this side during the last two days.

On the rocky ground of this region the tubes of *Eunice tubifex* were plentiful together with many sponges and gorgonoids.

Siphonochalina communis, Axinella donnani, A. tubulata, Isodictya sp., *Suberites inconstans*, etc, were met with, also *Juncella juncea* (in quantity) and *Gorgonia miniacea.*

Neither chanks nor *Pinna* were found.

The sand met with was, as usual on this coast, fine grained and largely calcareous, made up in the main of minutely comminuted shells. The quartz grains present were all extremely minute ; foraminifera were fairly abundant

In the afternoon as wind and sea increased rapidly we ran for shelter to the north-east side of Uppu Tanni Tivu By 4 p m. a very nasty cross sea got up and with the wind blowing half a gale the ship rolled unpleasantly at her anchorage. Towards sunset a strong land wind set in, the sky over the land murky-red and threatening

In the morning (May 14th) the landward side of the vessel, and of the awnings, funnel, stanchions, etc., was covered with a thick coating of impalpable red dust ; the murkiness over the land of the preceding night had been due to dust clouds, which, as they prevail throughout the south-west monsoon on this coast, must therefore form no inconsiderable factor in the production of muddy deposits in the sea

KUMULAM PĀR —Leaving the Uppu Tanni Tivu anchorage at 6-20 a m we steamed south to the small Kumulam Pār. A series of dives here in six fathoms showed the bottom to be rocky, some imes fairly clear of sand, at other times covered with from one to two inches   A few Holothurians and sponge fragments were brought up ; no trace of oysters was found

Steaming a quarter of a mile further another series of dives gave similar results ; flat rock with sand filling the depressions and with occasional small, loose fragments of calcrete upon the surface

At both stations the sand contained a large proportion of mud, so much that I have no hesitation in condemning this ground as hopeless and utterly unfit to rear oysters to maturity.   It may with safety be ignored in future inspections

The ground lying between the Kumulam and Vembar Periya Pars was next examined and found to consist of a coarser sand than any seen so far   Many medium sized quartz grains were present and though the water here was discoloured, no mud was actually found in the sand

Depth 10 fathoms   Temperature of sea at 4-30 p m 89° F. Specific gravity 1,022 80

VEMBAR PERIYA PĀR.—Continuing our course we arrived at 8-15 a m at a point which Captain Carlyon believed to be the north-east end of the Vembar Periya Par.

The steamer and the four inspection boats were employed in the examination of it for the rest of the morning

This par as outlined on the chart is of considerable extent, 3 miles in length by 1½ mile in breadth, the bank ranking next in size to the Tolayiram Par   The depth shown on the chart is 6 to 7 fathoms.

The examination proved hardly satisfactory, as out of upwards of 160 dives taken over an area of three miles long by one mile broad, but eight dives (in 8 fathoms) were on rock   All the soundings were between 6½ and 9 fathoms, the great majority being 8 to 8½ fathoms

On the small area of rock found by No 2 boat two small patches of pearl oysters—4 to 5 months old —were discovered, together with a large number of Suran (*Modiola barbata*)   One small fragment of rock bore a densely packed cluster of 10 individuals   Eunicid tubes and zoophytes were also present in considerable abundance.

The greater part of the sand was of a character approximating closely to that found on the Ceylon banks—bright clean yellow in general colour with plenty of quartz, cleaner and better sand even than that taken between this region and the Kumulam Par   Two dives gave tenacious black mud similar to that taken between Uppu Tanni Tivu and Nalla Tanni Tivu Pārs.

The pār is all a more or less pure limestone, mostly of fine grain.

Viewed by the nature of the bottom and the depth of water, I am very doubtful if the locality inspected is the Vembar Periya Pār; I believe the ground examined in reality lies seaward of the pār so named This Vembar Periya bank at all times must be difficult to find from want of good land-marks; there are indeed no charted land-marks observable from this position at sea, though several remarkable trees and clumps can be seen and might be utilized for lack of something better. To do this sketches of the relative positions of these trees and clumps should be made from different bearings and the positions of the principal objects marked accurately on the chart

Fortunately this is the only bank in any way difficult to locate, all others that are equally far from land being within sight of such conspicuous land-marks as beacons, lighthouses and pagodas.

On the evening of May 14th we returned to Tuticorin, the unsettled and threatening character of the weather making it doubtful if we should be able to do any further work at sea

Pending a decision upon this question, I occupied part of this stay ashore in endeavouring to locate any bed of window-pane oysters (*Placuna placenta*) that there might be in the neighbourhood. I had found prior to this several young individuals thrown up by the tide along the shore to the south of the town, and from the muddy character of the bottom I thought it useful to investigate further. Accordingly taking two fishermen I proceeded south along the coast and passed the salt factory scrutinizing as we proceeded the eroded edge of the low sandy land on the one side and the face of the littoral on the other. Along the shore of the bay-like estuary south of the salt factory large quantities of dead *Placuna* shells are observable in two places accumulated in dense masses and embedded in the sand some little distance below high-water mark. The shells appear as densely packed as those in the great heaps which mark the sites of former fisheries of this shell-fish along the shores of Lake Tampalakam near Trincomali in Ceylon.

Close by I saw others embedded in the adjacent sand hummocks, showing up wherever a section was exposed. In some places a depth of sand of fully 18 inches lay upon the shells. Meanwhile the fishermen had been wading about in the shallows and after trying various places found a small bed of the living animal and brought ashore a considerable number The majority of them were fairly well grown and approaching maturity The size of six typical individuals averaged 15½ centimetres in diameter, being almost perfectly circular in outline.

The men also reported large numbers of dead shells which however, probably belong to past generations.

Placuna lives well out of water for some time if kept in a cool situation, a property due to the faculty this shell-fish possesses of closing the valves tightly at all points round the margin, after the fashion of the edible oyster. In this particular case, several individuals which I put on one side for this experiment were found alive and vigorous thirty hours later, in spite of being drained of water and the temperature as high as 96° F in the shade.

In Ceylon a fishery of this shell was frequently leased by Government with considerable profit during the past century, the locality being an extensive shallow muddy bay on the north-east coast, close to the harbour of Trincomali.

In Tuticorin I could glean no information as to whether a fishery had ever been held in the neighbourhood, in view of the presence of living individuals and of the great piles of embedded shells along the sea-shore it might prove to be of advantage to Government if further search were made in suitable localities- backwaters, shallow estuaries, etc.—along the coast of the Madras Presidency with a view to locate any beds sufficiently large to provide a fishery and to ascertain also if pearls be sufficiently numerous to make such a fishery profitable I notice that Mr Edgar Thurston* records *Placuna* from Pulicat lake and Buckingham canal The former of these has, I understand, an area far exceeding that of Tampalakam Bay and should, therefore, receive special attention †

The weather had now improved temporarily and it was decided to spend two more days upon the banks, in order if possible to provide me with an opportunity to make a diving descent upon the Tolayiram Pär, a matter which I considered to be of great importance

---

* "Notes on the Pearl and Chank Fisheries and Marine Fauna of the Gulf of Manaar," Madras, 1890, page 27.

† Subsequent examination showed that these localities are not sufficiently open to the sea to support large beds The occasional individuals found occur only adjacent to the seaward openings or "bars"

Leaving Tuticorin on the morning of May 17th, we proceeded to a position south-west of Mela Onbadu Pār, where we spent the whole of the available time in taking numerous hauls of the dredge in $9\frac{1}{2}$ to 10 fathoms. My intention had been to select a locality which was a known chank bed in order to experiment with the dredge. I had accordingly requested Captain Carlvon to arrange to be taken to a suitable bank, the consequence being that the pilotage of the steamer was entrusted to a Parmandadai, who was, I understand, informed of my wish in the matter. The results showed that he deliberately placed the ship where I am absolutely convinced no chank bed exists.

This action, accidental or intentional, barred my way to demonstrate, as I had wished, the utility of the dredge in fishing for chanks. Fortunately one of the pearl banks fished in February 1905, on the Ceylon side, adjoins a chank bed, and having the use of the dredging steamer "*Violet*" at that time, I was able to satisfy myself by actual experiment that the dredge is an efficient implement for successful chank fishing.

The bottom where the Parmandadai took the "Margarita" was quite unlike that of any of the undoubted chank beds we had previously visited, while the depth of water, almost ten fathoms, was greater than that upon typical chank beds.

The ground was extremely rich in life; elongated cylindrical actinians of two species were abundant embedded in the sand, together with large numbers of an elongated Molgulid, which appears to live upright with the aboral extremity implanted in the sand. A small *Flabellum* sp. was plentiful on the surface together with a drab and grey *Pennatula* and many *Virgularia juncca*; arborescent rosy tipped Pennatulids were also characteristic, while several specimens of a hollow-stemmed coarsely-branched Alcyonarian (*Solenocaulon*) were taken. The latter were pink and white in colour and were accompanied in each case by a pair of small crabs and a pair of small Galatheids similarly coloured and obviously commensals; with some was a small commensal pink Gobioid sheltering in the hollow stem.* Some colonies were more uniformly suffused pink than others and on these the commensals were more uniformly tinted pink. On one colony, where white colouring largely prevailed, and where the margins of

---

* Described in the *Jul. Bombay Nat. Hist. Society*, September 1922, as a new species, *Pleurosicya annandalei*, Hornell and Fowler.

the branches alone were coloured pink, the crabs were uniformly white except for two splashes of pale pink on the anterior edge of the carapace. The large fan-shaped green alga with bulbous base embedded in the sand was the only alga found. A single specimen of *Lingula* sp. was one of the noteworthy acquisitions made in this locality

The sand was fine, clean, and with no trace of the mud which is a characteristic component and essential attribute of a prolific chank bed

Hence we moved north to the Tolayiram Pār, where we anchored after verifying the locality by cross bearings and by some trial dives, which indicated the presence of numerous young oysters

During the evening the crew caught a large number of trigger fishes (*Balistes*) and of strong-toothed fishes of the genus *Lethrinus*, the latter being known as Vellamīn,* among the Tamil fishers. The stomach contents of the *Balistes* were in this case free from incriminatory evidence in respect to pearl oysters, due probably to the fact that all the individuals were of a small size. In the case of the Vellamīn on the contrary the stomachs were crammed with fragments of pearl oyster shells upon which these fishes appeared to be feeding exclusively. As six of these, all of large size, were caught within two hours, it is plain that these fish are now inflicting enormous havoc upon the bed of oysters on this pār

The next morning I donned the diving dress, and the water being fairly clear I had an excellent opportunity of examining the bank

The depth where the descent was made is 9½ fathoms. I found the bottom very variable; in most places a covering of about an inch and a half of sand lay upon flat-surfaced rock. Here and there the rock protrudes or lies level with the general sandy surface These exposed patches are small in area, usually from 1½ to 2 feet in diameter. A limited amount of small cultch is present on the surface to which the majority of the oysters adhered. The cultch consists of short fragments of much worn branches of coral ("challai"), quite small Nullipore balls (*Lithothamnion*), the tests of dead echinoids (chiefly of *Clypeaster humilis*), fragments of calcrete and such like

The oysters were fairly abundant, their numbers obviously curtailed by the quantitative limitation of the cultch. They appeared

---

* *Lethrinus nebulosus* and other species

to be of two generations, the one four to five months, the other ranging from six weeks to two months

Sponges were not obtrusive ; several specimens of the peculiarly massive *Petrosia testudinaria* were seen but the commonest species was one which interiorly is yellow, while the exterior is more or less tinted with pale green   In its growth it envelopes much sand in its substance, which is generally level with the sand surface, sending up stout tapering branches at intervals Its general appearance is inconspicuous.

No living coral was seen except a small *Favia* sp and an occasional Gorgonid

While on the bottom I saw many fishes including both *Kilati* (Trigger fishes) and *Vellamin* together with a beautifully stripped sponge-eating fish (*Holocanthus imperator*) The Asterid, *Pentaceros lincki*, was not abundant. No Suran (*Modiola barbata*) was seen

After I had completed my examination of the bottom, Captain Carlyon being of opinion that it was now too late in the season to do further work, and as the coal supply was on the point of giving out, under the circumstances I determined to bring the investigation to a close.

When the anchor was brought up we had a repetition of our former experience on this bank—the chain being studded with young oysters of about two months old affixed by their byssal cables near the attachment to the anchor

Two hours later we landed for the last time at Tuticorin and although I had not been able to examine every bank charted, I considered that I now knew the general characteristics of the principal groups sufficiently well to answer all practical purposes and enable me to furnish solutions to the majority of the problems I had set out to solve.

VII.—COMPARISON OF THE CHARACTERISTICS AND RELATIVE IMPORTANCE OF THE VARIOUS PEARL BANKS

In the present section an attempt is made to seize upon the essentially characteristic features of the chief pārs—historical,

topographical, physical and biological—and therefrom by comparison with one another and with typical banks on the Ceylon side
to evolve a knowledge of their relative economic importance in
regard to the prospect they respectively hold out of successfully
maturing such oyster spat as may from time to time settle thereon
Such a comparative survey will also go some distance towards
enabling us to say what direction any measures of cultivation
should take, if it be found advisable or possible to assist nature by
artificial means

The configuration of the Indian coast of the Gulf of Mannar is
simpler than that on the Ceylon side   On the former there is no
great shoal like that of the Ceylon Karativu, which stretches northwards into the sea for a distance of nine miles giving a certain
amount of shelter to a great area of varied bottom, rock and sand,
lying in the Bay of Kondachchi   On the contrary the Indian pearl
banks lie open to the full force of the south-west monsoon which
on this coast sweeps up in great violence from south to north.
Again lying as they do on the west side of the Gulf, they also
experience much rough weather during the north-east monsoon, a
time when the Ceylon banks, lying under the lee of the land, enjoy
comparative quietude   The period of immunity from storm disturbance on the Indian coast is accordingly greatly curtailed and is
restricted under normal conditions to the months of February,
March and April Occasionally fairly quiet conditions prevail
during the greater part of May—the onset of the south-west monsoon in full force being experienced somewhat more tardily there
than on the Ceylon side

This geographical disability of the Indian banks is linked with
and intensified by the mechanical disadvantage entailed by the
inferior character of the sand on that side, its finer grain and the
admixture with it of mud—characteristics which contribute to
increase greatly the turbidity of the water whenever heavy seas
sweep the pearl bank region   As already noted, these are conditions which have probably become intensified concurrently with
the erosion of the southern extremity of India and which tend,
though with extreme slowness, in the historic sense, to reduce the
pearl oyster productiveness of this locality—deductions from which
we infer greater prosperity in times past   That there was such
anterior prosperity we have indications in the existence of remains
of ancient oyster-shell heaps close to Cape Comorin, in the frequent

allusions of classical writers to the wealth of the pearl fisheries of Kolkoi which town we have seen was situated on the river Tambraparni, in the statement of Friar Jordanus that as many as 8,000 boats were engaged about the year 1330 in the Indian and Ceylon fisheries and in the fact that Kayal or Cail near Pinnakayal is spoken of by Marco Polo, Ludovico de Vathema, Barbosa and other mediaeval travellers as the headquarters of the pearl fishery —a "great and noble city inhabited by jewellers who trade in pearls."

Dealing with the conditions as they are at present, we find that so far as our available modern records permit us to judge, all the known oyster-productive banks are comprised in the division which I have termed the central, lying between Vaippar and Manappad point, a distance roughly of 40 miles

A list of these banks and of the groups into which I propose to classify them has been given above on pages 51 to 53

It now remains to describe the varying characteristics of each group and to institute comparisons as detailed as the material at our command will allow.

I. TOLAYIRAM GROUP

This group possesses the distinction of being the most productive and remunerative collection of pars along the Tinnevelly coast. Two banks only are comprised under the group title—the large Tolayiram Par and the small Kutadiai Par The former is by far the largest of the productive pars ; the latter, which lies to the south-west extremity of its huge neighbour, being on the contrary one of the smallest, an oval-outlined rocky patch one mile long by half that in breadth

The Tolayiram Par lies 8 to 11 miles off the coast and opposite Hare Island and Tuticorin Bay, the northern extremity due east from the town of Tuticorin In shape the bank is roughly crescentic, the concave side turned shorewards Its long axis lies roughly north-east by south-west measuring over six miles in this direction. The width varies from one to two miles, broadening as we approach the upper extremity The depth of water over it ranges from 8 to 11 fathoms.

Nine fisheries have taken place upon this locality during the past 120 years, namely, in 1784, 1787, 1807, 1810, 1822, 1830, 1889, 1890 and 1908

The annexed table shows the results of these fisheries so far as I can obtain particulars :—

Year				Number of oysters fished	Gross Govern- ment revenue	Net Govern- ment revenue
					Rs,	Rs.
1784	...	...			42,420	42.420
1787	..	.		.. ...	63,000	63,000
1807	..	...	...	71,647,305	2,91,539	.. ...
				More than		
1810	..	...	..	22,000,000	2,38,897	... ...
1822	..	..	..		1,55,693	... ..
1830	..	..	.	.. .	Separate revenue not given	...
1889	...	..	..	..	12,600,531	1,89,984	1,58,483
1890			.	1,806,762	25,061	7,803
1908	...	..	...	1,101,642	10,218	7,282
				Total revenue over Rs 10,16,812		

The name Tolayiram Pār, literally "900 banks," pithily describes the peculiar physical conditions which prevail over the area so denominated. The character of the bottom is an alternation of rocky patches scattered irregularly in a vast setting of sard.

The sizes of the outcrops of rock differ greatly, from little tabular fragments a foot or two across to great areas of several acres in extent. The sand is nowhere deep, seldom forming a layer of more than six inches in depth, filling up inequalities in the rocky framework of the bank

The rock is a fine-grained limestone compact and resonant, the colour yellowish brown. Here and there a small admixture of quartz is present but never in any large proportion. Loose blocks and many parts of the exposed surfaces are in a "rotten" condition, tunnelled and excavated by boring molluscs and occasionally by *Cliona*.

The character of the sand is fine-grained and almost entirely calcareous—a similar material to that from which the underlying rock has originated.

Cultch is fairly abundant in places, scattered over the sand. It consists of dead shells, broken branches of Madrepore coral ("challai"), Echinoid tests and similar material.

A striking parallelism can be traced in nearly every character-istic between this bank and the well-known Periya Pār on the Ceylon side of the Gulf of Mannar. In both cases the bottom consists of a few inches of sand covering flat rock in those places where the rock does not outcrop and with a fair amount of small cultch scattered over the sand

The average depth, 9½ to 10 and 11 fathoms, is the same in both ; both are situated further seaward than any other true oyster pars on their respective sides—the Periya Par 16 to 18 miles off land, the Tolayiram Par 9 to 11 miles

The faunistic characters approximate in a remarkable manner; the larger and more conspicuous species of animals are the same in both localities.

In the sponge, coral, echinoderm, molluscan and fish fauna there is practical identity. One list will serve for both.

Thus we have as common to each :—

Petrosia testudinaria, Spongionella nigra, and *Axinella donnani* typifying the sponges ; the abundance on these two banks of *Petrosia* is one of the most remarkable of the many of the striking resemblances between these banks, for this sponge, striking in its strangely massive form, may be said to be limited to them. I have scarcely ever seen it elsewhere in profusion.

Corals are scarce on both banks, represented by isolated colonies of Astraeids (*Favia* sp.) and of *Meandrina.*

Occasional Alcyonarians and Pennatulids are found, together with numbers of knobbed horse chanks and small lamellibranchs of identical species   *Pentaceros lincki, Linckia laevigata* and *Antedon* are the chief starfishes on the Tolayiram Par

On both banks the fish population as represented by the trigger fishes (*Kilati*), and Vellamin (*Lethrinus* sp.) and gobies, appears to be greater in numbers than on the banks nearer the shore.

On the whole both banks are decidedly poor faunistically, with little diversity of life-forms which in the majority of cases are also poor numerically. The absence of Madrepores, of *Pinna* and of the tubes of *Eunice tubifex,* is characteristic and noteworthy.

The Periya Par is cited by Professor Herdman [*] as especially suitable for dredging over. The Tolayiram Par is equally so, or if anything somewhat superior as the rocky surface is quite free from upstanding growths or rugged inequalities

Reviewing the history of the Tolayiram Par as shown by the inspection diaries dating from 1860, we find that subsequent to the year named—when the bank was covered with oysters said to be 3½ years old—there are records of the bank having been stocked

---

* *Loc. cit,* Pt. 1, page 111

extensively with spat four times, one of which resulted in the fisheries of 1889 and 1890

The particulars of these are—

1863 "Some young oysters" (It would appear that the numbers could not have been large Suran was noted as present the same year)

1874 "Some young oysters on Kutadiar Par with Suran"

1878. "Thickly stocked with oysters of one year age"

1881. "Some oysters of one year"

1884 The inspection summary reads "Plenty of oysters of one year age; clean and healthy." These oysters survived and furnished the two successive fisheries of 1889 and 1890, at which a total of 14.407,293 oysters were fished

1904 Since 1890 only one spat-fall has been recorded—that found during the present year's examination

The bank was not examined in the spring of 1861, nor in 1862, 1864, 1868, 1870 1874, 1893 and 1900

From the above we observe that out of a total of 44 years, 1860—1904, there have been five recorded spat-falls on this bank with the probability of a sixth in 1874, when this bank was not examined although the adjoining Kutadiar Par bore young.

The bank brought one lot of these—that of 1884—to maturity and from what I can see the prospects of a fishery resulting from the present population of young oysters are good if they survive till next spring By that time they will be too large to suffer much from the depredations of oyster-eating fish (Trigger-fishes and Vellamin). They will then be more robust and better fitted to endure the discomforts and danger of starvation, which are the concomitants of the disturbed water conditions during the stormy period of the year.

Comparing the history of the Periya Par, we find that in 26 years ending 1904, this bank was restocked at least twelve times without yielding a fishery. We know also that one fishery, that of 1879, is the only one yielded by this bank during the past century (7,645,901 oysters realizing Rs. 95,694).

So while the Ceylon bank is infinitely more fertile in the number of times it is replenished with oyster spat, its Indian counterpart has greater reliability, six times do we know that it has brought its oysters to fishing maturity, namely, in 1784, 1787, 1807, 1810, 1822 and 1889-1890, and very probably a third time as well, for the oysters noted in 1860 as 3½ years old were in all probability there in 1861, in the spring of which no examination was made, the officers in charge being busy with the fishing of oysters of a similar age on the inshore pars This divergence in results is due in great part to the Ceylon bank being situated in a relatively more exposed position being close to the edge of the precipitous submarine cliff that margins the seaward aspect of the Ceylon Pearl Bank plateau. As a consequence the heavy seas which characterize the period of the south-west monsoon break in unmitigated violence upon the Periya Pār, whereas on the Indian coast the movement of the water during the same season has undergone considerable amelioration when it reaches the Tolayiram Pār from travelling over a couple of hundred miles of comparatively shallow water.

Faunistic and many physical (chiefly geological) characteristics link the Tolayiram with the Periya Pār, but in regard to the aspect and degree in which the former meets the fury of the south-west monsoon, its position is more comparable with that of the Cheval Pār which lies on the leeward side of the Periya Pār, and is as consistently reliable as the latter is the converse

The value of the Tolayiram Pār may be assessed as midway between the Cheval Pār and the Periya Pār, inferior to the former chiefly by reason of oyster spat being less abundant and to current conditions (surface-drift) being less favourable to the deposit of such spat on the Indian than the Ceylon side ;—partly also to the conditions of life being somewhat less favourable on the Tolayiram Pār owing to the greater amount of sediment present in the sea on the Indian side

The data for the institution of comparison between the rate of growth normally characteristic of the Tolayiram Pār oysters with that of oysters from typical localities on the Ceylon side rest upon a single series of measurements and weights of that generation of the former that survived to a fishable age in 1889 The resultant comparisons based upon these dimensions are highly instructive and while in my opinion I believe it is probable that they are quite typical of the normal progress of growth of oysters on this pār, further series of growth observations are desirable and inspectors after this should be instructed to record the necessary particulars on every available opportunity.

Two methods of comparison are available : (a) the external dimensions of the oysters when alive, and (b) the weight of the cleaned shells.

Making use of the former method we find usually but little increase in the length and depth of the shell after the third year, the shell-secreting energy of the animals being thereafter occupied chiefly in adding to the thickness of the valves

I now attach greater importance to observations upon the average weight of oyster shells than upon measurements of length and depth, increase being nearly as steadily progressive in old age in the case of weight of shell as it is during the first three years of existence; it furnishes us with the most reliable guide available in the assessment of age that I know of.

But we need to have considerable knowledge of the special growth peculiarities of the ground we deal with. Some pārs by reason of abundant food supply hasten the growth of their oysters to a surprising degree, while others where less favourable conditions prevail bear oysters of an unhealthy appearance and of stunted size.

On the Ceylon side two distinct types of oysters are found, the one large and vigorous, peculiar to the Southern and Eastern Cheval and Moderagam Pārs, the other slow-growing, small and stunted, characteristic of the rocky banks of the Muttuvaratu, Mid-West and North-West Cheval.

We will now proceed to compare the sizes and weights of the generation of oysters carefully guarded on the Tolayiram Pār by Captain Phipps from 1884 to 1889 with those of oysters of the two types referred to on the Ceylon side

Weight of Pearl Oyster shells

Locality	Date	Age	Number weighed	Weight per 100 pairs of valves	Increase in weight in year preceding.
		YEAR	PAIRS	OUNCES	OUNCES
	March 1884	¼	10	1 00	
	October 1884	1¼	10	3 75	
	March 1885	1¾	10	6 25	5 25
	October 1885	2¼	10	7 00	3 25
	April 1886	2½	10	7 50	1 25
Tolayiram Par	November 1886	3¼	10	8 50	1 50
	March 1887	3¾	10	10 75	3 25
	October 1887	4¼	10	13 00	4 50
	November 1888	5¼	10	15 25	2·25
	(Fished) March 1889	5¾	10	16 62	
	,, ,, 1890	6¼	3	17 64	1 02

Weight of Pearl Oyster shells

Locality.	Date.	Age	Number weighed.	Weight per 100 pairs of valves.	Increase in weight in year preceding.
		YEARS	PAIRS	OUNCES	OUNCES.
Western Cheval Pār * (Type intermediate between the freely grown oysters of the South and South East Cheval and the extremely stunted ones from the Muttuvaratu.)	March 1871	¾	13	2·50	
	„ 1872	1¾	13	7·50	5·00
	„ 1873	2¾	50	11·88	4·33
	November 1873	3½	100	12·81	
	(Fished) March 1874	3¾	45	15·31	4·43
	„ „ 1875	4¾	47	18·75	3·44
South East Cheval Par * (Free growth type.)	March 1874	¾	33	2·25	
	„ 1875	1¾	60	8·50	6·25
	„ 1876	2¾	150	13·44	3·94
	„ 1877	3¾	51	18·75	5·31

* In Captain Donnan's table the ages of these are given as three months older than shown here, the figures now given, I believe, approximate more closely to the actual ages.

Analysing the above table we obtain the following comparisons :—

Tolayiram Pār.

		OZ
Weight at	¾ year ...	1·00
„	1¾ years .	6·25
„	2¾ „ .	7·50
„	3¾ „	10·75
„	4⅓ „	13·00
„	5¾ „	16·62

North-West Cheval Pār

Weight at	¾ year ...	2·50
„	1¾ years ...	7·50
„	2¾ „ ..	11·88
„	3¾ „	15·31
„	4¾ „ .	18·75

South-West Cheval Pār.

Weight at	¾ year ...	2·25
„	1¾ years ...	8·50
„	2¾ „	13·44
„	3¾ „	18·75
„	4¾ „	..

13

The weight of three typical oysters from the fishery held on the Muttuvaratu Pār in 1891 is 5.375 ounces, equivalent to a weight of 17.916 ounces per 10 pairs of shells when the oysters were approximately 6½ years old.

The annual increase in weight of oysters on the Tolayiram Pār compares, as will be seen from the above, very unfavourably with the comparatively stunted oysters characteristic of the Western Cheval. The nearest approximation was at the age of 1¾ years when 10 Tolayiram Pār oysters weighing 6¼ ounces were but 1¼ ounce less than the weight of a similar number from the North-Western Cheval. During the next twelve months, however, the latter gained 4½ ounces as against an increase of 1¼ ounce by the former. This disparity continued to increase more slowly thereafter, but unfortunately for want of data we cannot give the exact amount for the age of 4¾ years.

It is a pity that we have not available a record of the yearly weight increase of the oysters fished on the Ceylon Muttuvaratu Pār in the same years as those of the Tolayiram Pār. If we had, I think we should find that there would be shown close approximation between the two; the Muttuvaratu oysters of that generation were markedly stunted and poor and the fishery of 1889 was decided upon only after considerable hesitation.

The only datum I possess is the weight given above of three typical oysters from the 1891 fishery. This which is equivalent to a weight of 17.916 ounces for 10 shells at 6½ years of age as against 17·64 ounces for a similar number of 6¾ years old oysters from the Tolayiram Pār in 1890 indicates practical identity in growth-rate.

By the courtesy of Captain Carlyon I have been enabled to measure a few individuals of this last fished generation of Tolayiram Pār oysters and append a table thereof in which the measurements of some of the oysters from the Muttuvaratu Pār are included for the sake of comparison. The numbers are too restricted to give an average that may be taken as thoroughly trustworthy. They constitute, however, the only data available and till systematic records extending over a considerable series of years be obtained by work in the future it is well to place them on record.

Date	Age	Size	Average.
March 1885	1¾ years	40 × 64 × 23 millimeters 60 × 53 × 22½　　,,	65 × 58 5 × 22 75
April 1886	2¾ years	62 × 55 × 23½　　,, 63 × 58 × 27½　　,, 67 × 64 × 31　　,, 62 × 62 × 30　　,,	63 5 × 70 5 × 27 5
October 1887	4½ years	71 × 70 × 32　　,, 79 × 70 × 32　　,, 80 × 74 × 35　　,	76 66 × 71 33 × 33
November 1888	5⅓ years	77 × 80 × 32　　,, 78 ⅄ 78 × 33　　,, 78 × 74 ⅄ 33½　　,	76 66 × 77 33 × 32 83
March 1889	5¾ years	78 × 80 × 33　　,, 80 × 74 × 31　　,,	79 × 77 × 31 33
,,　　1890	6¾ years	72 × 68 × 31　　,, 70 × 73 × 34　　,, 76 × 72 × 31½　　,,	72 66 × 71 × 32 16

Muttuvaritu Pār, Ceylon

March 1891	6½ years (deep short oysters covered with living growth —Lithothamnion and corals ; much corroded by the tunnelling of *Cliona*).	73 × 63 × 36 millimeters 67 × 57 × 35½　　,, 80 × 58 × 36　　,,	73·33 × 59 33 × 35 83

It would appear from the preceding tables that the growth of the Indian oysters is distinctly retarded after the third year, the life conditions being more favourable to the young than to the old — a condition which I believe will be found due largely to the great abundance of encrusting organisms, sponges and polyzoa especially, which begin to flourish upon the valves of the Indian oysters in wonderful abundance from the age of 1½ year　A similar state of marked retardation in growth is characteristic of the oysters from the Ceylon South Moderagam Pār, after the attainment of the same age, a retardation coincident with the appearance of luxuriant sponge tunicate, and polyzoa growth upon the valves　On the South Cheval where such commensal growth is rare, no such marked slackening in the rate of growth is apparent.*

The oysters of the Tolayiram Pār in October 1887, when 4⅓ years old, gave a pearl valuation of but Rs 3-11-5 per 1,000 and as

---

* See my ‘ Report on the Biological Results of the Ceylon Pearl Oyster Fishery of 1904 ”—*Ceylon Marine Biological Reports*, No. 1, Colombo, 1905.

this was much too low to justify a profitable fishery, it was not till after the valuation of November 1888, affording a valuation of Rs 13-12-8 per 1,000, that a fishery was decided upon. The oysters were therefore 5¾ years old when first fished in 1889

Against this we find that the comparatively stunted oysters of the North-West Cheval Pār were ready for fishing at 3¾ years of age—a sample lifted in February 1874 giving the high valuation of Rs 36-8-0 per 1,0c0

Finely grown oysters on the South-East Cheval were also fished in 1878 at the reputed age of 4¾ years ; their valuation three months prior thereto was Rs 39 14-2 per 1,000

The oysters fished on the Muttuvaratu Pār in March 1889 were reputed to be 4½ years old, and in the November preceding, at the approximate age of 4⅙ years, the valuation sample worked out at Rs 10-2-4 per 1,000

2 UTI PĀR GROUP

A chain of six banks, the Nagara, Uti, Uduruvi, Kilati, Attuvaiarpagam and Patarai Pārs constitute this group. All are of small and of about equal size, averaging from ½ to ¾ of a mile in diameter. They lie in a depth of 7 to 8½ fathoms, landwards of the Tolayiram Pār, at a distance of 5 to 7 miles from the shore. They stretch north and south about 3 miles

The area is essentially rocky, the proportion of sandy ground intermingled with the rock insignificant

Faunistically this area is richer and more diversified than the Tolayiram region, the intimate intermingling of rock and sand upon the latter producing effects when the sea is disturbed which but comparatively few species of animals can tolerate.

The fauna agrees closely with that of those Southern Ceylon banks lying off Negombo, notably with Uluwitte Pār which lies at the same depth

The features characterizing the Uti banks in common with those off Negombo are as follows :—

An abundance of sponges including a larger number of small species than in the case of the Tolayiram. *Siphonochalina communis* with its numerous commensals is amongst the most common ; fixed corals are scarce, Zoophytes are profuse with many colonial masses of *Filigrana* tubes and everywhere the curious branched tubes of *Eunice tubifex*. *Pinna* sp covered with large Balani are conspicuous on the sandy ground

The rocky bottom on the Uti banks is calcrete, containing in some places a considerable quantity of quartz grains embedded in a calcerous matrix—a quartzose limestone.

The only fisheries recorded from these banks during the last 120 years took place in 1792, 1830 and 1860-1861. In the last instance the oysters, said to be 4½ years old, were abundant and of good fishing value. Adjacent banks are usually fished in the same season with these and separate figures of the number of oysters fished from the Uti banks are not available

3. PASI GROUP.

Two banks, the Pasi Pār and the Attonbadu Pār, may be linked together under this head. They lie 6 to 7 miles off Hare Island, Tuticorin, at a depth of 8 to 9 fathoms, and are situated nearly midway between the Uti Pār group and the western margin of the Tolayiram Pār.

These also were fished in 1861 together with the adjoining Uti Pār and associated banks. Since then the only records of oysters present in quantity are—

1863	...	... "Very young oysters and Suran."
1876	...	.. " Plenty of young oysters of 1½ years."
1881	...	... "Large numbers of oysters of one year of age with Suran in some places and covered with weeds."

In addition, in 1894, 1896 and 1901 some few young were found, but as their number was limited we may disregard them and draw the inference that like so many other banks on this side these two suffer rather from a shortage of spat than from inability to support in health those that do appear and survive the dangers of the first 18 months of existence.

The bottom on the rocky patches is the usual calcrete, the remainder of the ground fine sand with occasional chanks.

The young oysters found in 1901 lay principally on the sandy stretches

4. CRUXIAN GROUP.

Another group of small pārs, three in number, lying west of the island of Vantivu and about six miles from the mainland The three constituent pārs, Cruxian, Tundu and Vantivu Arupagam, are to the north-north-west of the Uti group in rather shallower water, 6 to 6½ fathoms.

The bottom on the pars consists of level stretches of continuous rock, brownish tinted calcrete exactly similar to that on the Uti Pars.

The fauna differs considerably from that of the last-named banks. Sponges are less extensive, *Siphonochalina communis* being the most conspicuous and numerous.

Among other animals noted were large *Pinna* sp. in abundance rooted in the thin layer of sand covering the rock in many places, with *Balanus* and zoophytes crowding the exposed surfaces of the *Pinna*, *Eunice tubifex* in quantity, Heteroneid form of *Nereis* sp. in the canal system of *Suberites inconstans*; *Botrylloides* sp ; *Turbinella pirum* in the sand on the western side.

The large fishery of 1861 was contributed to from these banks which appear more favourably situated than many others for receiving spat falls, some eight being recorded since 1861. Unfortunately in only three instances, 1878, 1884 and 1902, did the re-stocking take place on an extensive scale—even in 1902 the quantity of 1½ to 2 years old then present was estimated at but 1,700,000, a number too small to give good results two to three years after in view of the unpreventable wastage that must be allowed for.

In many respects the Cruxian group has points of resemblance with the North and South Moderagam Pars on the Ceylon side, notably in the in-shore situation, the comparative shallowness of the water and in the characteristic abundance and association together of *Pinna* and *Balanus*.

The ground referred to on the Ceylon side is much the more clean of the two, both faunistically and physically ; the sand there is of the usual coarse grit and this, by the attrition of its movement during disturbed weather conditions effectually scours the bank, keeping down the growth of weed and other organisms unprotected by a hard external protective casing.

This mechanical cleansing of the bottom is nowhere well seen on the in-shore Indian banks where the fineness and low specific gravity of the sand lacks not only an adequate scouring force, but by reason of the presence in it of a certain amount of mud exercises a retarding influence upon oysters when they are present—an influence resulting in a stunting of the growth.

The fact has long been noted [*] that the size of Ceylon oysters of a given age from the Cheval Pār is markedly superior to that of those of the same age from the in-shore Indian banks, the latter approximating more closely to those from the Muttuvaratu Pār, a bank with a bad reputation for the starved appearance characteristic of its oysters.

5. VAIPPAR KARAI GROUP.

The largest of these is the Vaippai Karai Pār, a bank of some importance not located upon the present inspection chart. From the observations made and the information supplied by the pār mandadai, it appears to lie north-west of the Devi Pār and about five miles due south of the village of Vaippar. The other banks in this grouping are the Devi, Pernandu, Padutta Marikan and Padutta Marikan Tundu Pārs, varying in diameter from half to three-quarters of a mile. Depth 6 to $6\frac{1}{2}$ fathoms.

The bottom is of the usual reddish-brown limestone common to the other groups in this neighbourhood, interrupted and more or less overlaid by a fine muddy sand, the larger particles consisting chiefly of comminuted shells. Numerous dead pearl oyster valves, entire and also fragmentary, were abundant, fully $1\frac{1}{2}$ year old; of live ones but a few odd individuals were found, greatly overgrown with tunicates and polyzoa and distinctly stunted in appearance.

The sand on the Vaippar Karai Pār is appreciably more dirty and muddy than that on the Cruxian Pārs, a difference due to the vicinity of the embouchure of the Vaippar river. The other pārs of the group are probably less affected, but all have borne mature oysters, the group being included in the fishery ground of 1861.

The faunistic characters approximate to those of the Cruxian Pārs. *Pinna* sp bearing large *Balani* predominate. A few corals (astræids) were seen with leptoclinids and zoophytes.

Sponges are neither numerous nor conspicuous.

It appears from the records that these banks have suffered neglect in recent years, which in view of the fishery held there in 1861 and of the record by Captain Phipps of an abundance of young oysters in 1867, 1873, 1877, 1881 and 1884 they do not justify

* Thomas, H, Sullivan, *loc cit*, page 14.

Thus the Karai Pār received no attention for the years 1887 to 1894 and again from 1897 to 1903, both inclusive, a period of eight years in the one case and of seven in the other

In the case of the other pars of the group the years of neglect are 1888 to 1890, 1892, 1893, 1898, 1900 and 1901, eight years in all

It is quite conceivable that fishable oysters were missed through such omission and it emphasizes the contention I make elsewhere for a reorganization of the work of inspection upon such a scientific basis of accuracy and method as will preclude such lengthy periods of neglect.

A significant incident pointing to the imperfection of the methods in use in the management of these banks is the statement made in Mr H. Sullivan Thomas' report * that oysters of 2½ to 3 years of age were found in December 1869 upon the Pernandu, Padutta Marikan and Padutta Marikan Tundu Pārs, while the entry for March 1869 states that these banks were totally devoid of oysters,—"blank" Comment is superfluous on such a state of affairs, not unknown either in the past history of the Ceylon banks †

The Padutta Marikan Tundu Pār was one of the banks fished in 1830, the only record of a fishery on this pār during the past century.

6. NENJURICHCHAN PĀR GROUP

Three of the usual small pārs, ¾ to one mile long, compose this group, namely, Nenjurichchan, Kundanjan and Mela Onbadu Pars and cannot be treated otherwise than as a single unit They lie at a distance of about 6 miles from the shore midway between Tuticorin and Pinnakayal The depth is 7¾ to 8¼ fathoms

The rocky surface is extensive and comparatively free from inorganic sand, what there is being composed largely of Foraminifera (*Orbitolites* and *Heterostegina*) The rock surface is level and well adapted for dredging purposes.

Physically and faunistically this group resembles closely the seaward side of the Ceylon Muttuvaratu Pār Like the latter it is rich in sponges and in Gorgonoids (*Gorgonia miniacea, Suberogorgia suberosa, Juncella juncea*), while the long-armed Asterid *Linckia laevigata* is fairly common The sandy region to the southward is remarkable for the occurrence there of *Lingula* and for the abundance of *Solenocaulon tortuosum*, and of *Pennatula* and *Virgularia*

---

* *Loc. cit.*, page 52

† Twynam, Sir William—"Report on the Pearl Fishery of 1888," Ceylon, 1888, page 13 , also Stewart—"Account of the Pearl Fisheries."

The group has a disappointing history well expressed in the name of the median par—Nenjurichchan, literally "Heart-harrower." Why this should be so is difficult to say as the group lies less than a mile to the south of the Tolayiram Par group; even the sandy stretch separating these groups carries occasional clusters of oysters and on the chank bed to the north-west it is not uncommon to find a dozen young oysters making use of the chanks in the absence of cultch and rock.

Probably the reason for such continued lack of oysters is due to some peculiarity in the set of the surface drift over these beds.

This group should receive regular attention during the next few years with a view to elucidate the reasons for this characteristic, note being taken (and recorded) of the character of the surface at each inspection, together with particulars of the relative abundance of the chief organisms met with, sponges, gorgonoids, corals, the tubes of Eunice, suran, chanks, fishes and seaweeds.

7. Puli Pundu Group

South-west of the Nenjurichchan group, this collection of small rocky banks comprising the Vada Onbadu, Saith Onbadu, Puli Pundu and Kanna Puli Pundu Pars, is situated about 9 miles north-east of Pinnakayal and some 8 miles west from the coast. The depth ranges between $7\frac{1}{2}$ to $8\frac{1}{2}$ fathoms.

The bottom of the pars is of flat-surfaced rock, somewhat patchy in distribution. Here and there is a small amount of cultch, more especially on the landward side, where a considerable amount of water-worn coral branches, "challai," is present.

The par is mostly a fine grained and exceedingly dense limestone, reddish brown in tint and so hard as to ring under the hammer. Occasionally the traces of dead massive corals, *Astræa* or *Meandrina*, appear embedded *in the surface layer of this rock*, and are usually much bored into by tunnelling molluscs and sponges.

The parchment-like tubes of *Eunice tubifex* are most profuse, their lower portions penetrating the tunnels already existing in the surface of the par-calcrete. The usual massive sponges, *Siphono-chalina communis, Spongella nigra*, and *Suberites inconstans* are met with, while off the edge of the banks on the west and north chanks were found in number together with an occasional *Pinna*.

The history of the group is disappointing, no record existing of any fishery having taken place here, although there were spat falls

14

noted in 1867, 1874, 1878, 1885, 1895, 1897 and 1901, all of small
extent and of no practical importance.

The banks were not examined during the 8 years between 1886 and 1895.

Fishes are very plentiful on this ground and the area of rocky
ground exposed is practically insignificant compared with the area
of sand, while cultch is quite insufficient   It is probable that in
these three disabilities we have the reasons for the smallness of
the numbers of oysters noticed here from time to time

8. INNER KUDAMUTTU GROUP

A series, stretching north and south, of 6 small banks lying 5 to
6 miles off the coast between Pinnakayal and Kayalpattanam.
The most northerly is the small Pinnakayal Seltan Par, the most
southerly a small bank, unnamed upon the chart, lying a quarter
of a mile south of the Saith Kudamuttu Par—the depth in all cases
being $7\frac{1}{2}$ to $8\frac{1}{2}$ fathoms

The general character of the rocky ground is almost identical
with that characterizing the Uti Par group which lies in the same
depth of water   Many of the larger organisms found in the latter
locality are also present here, sponges and Eunicid tubes coming
up at nearly every dive   *Pinna* and *Balanus* were noted as absent
from these banks—common features of the Uti pars   As on the
latter, a few odd oysters remain from the generation noted in 1902
as being from $1\frac{1}{2}$ to 2 years old; all were more or less enveloped
in the orange-red sponge *Clathria indica*

In 1818 Kudamuttu, Saith Kudamuttu and Pudu Pars gave a
fishery yielding Rs 1,67,693   Ten years later they were fished
again in conjunction with the neighbouring pars, and from an
entry in Captain Phipps' list * that oysters $2\frac{1}{2}$ to 3 years old were
present in May 1860 and that no inspection was made in the two
following years, I think there can be no doubt that mature fishable
oysters were here also in 1861 or 1862, not being fished owing to a
large number of other banks being stocked at the same date and
receiving preference in the order of fishing

The rocks show some diversity in character, dense and compact
limestone passing in some places into a somewhat quartzose stone
having a calcareous matrix   The hard bottom is much cut up by
more or less extensive stretches of sand   Here and there we meet

---

* Thomas, H Sullivan, *loc cit* , page 58

with loose fragments of calcrete similar in composition to the bed rock of the par; dead coral is fairly common in the form either of much honeycombed tabulæ or of rolled and much worn broken madrepore branches, derived probably by the action of backwash and under-current from the extensive coral reefs that fringe the adjacent coast.

Chank beds lie to the south, east, and west of these banks, forming virtually a girdling of chank-producing sands.

A list of the common forms of life met with here is given on page 74 together with other details

The term Kudamuttu used in the names of these banks is significant.   It means literally the "Pearl Bay," so that the shallow indentation off which these banks lie and which has Hare Island, Tuticorin, and Trichendur point as its northern and southern limits, with the mouth of the Tambraparni river at the centre of its curve, appears to have been termed the Pearl-bay *par excellence*, from the renown of the pearl fisheries held there Korkai and Kayal were successively at the embouchure of the Tambraparni, so we have in Kudamuttu further indirect evidence that the towns named were located near the centre of the most prolific pearl fisheries of early and mediæval times, the periods when they flourished respectively

9. OUTER KUDAMUTTU GROUP

This is a congery of some six small banks lying due east of the inner Kudamuttu group It measures some two miles north and south by the same from east to west, with an average depth of 9 to 10 fathoms.

No fishery is recorded from these banks; neither do we know of any extensive spat fall in any year since the inspection record begins in 1863.   Time did not permit of an extensive examination this year

10. KADIAN GROUP

This collection lies about seven miles west of Pinnakayal and due south of the Kudamuttu group from which it is separated by a narrow chank bed   To the south it marches with the Karuwal group In depth it agrees with the former—$7\frac{1}{2}$ to 8 fathoms

The two principal patches of rocky ground are the Kadian and Kanawa Pārs, each of about half a mile in diameter.   The whole group covers an extent measuring approximately two miles from north to south by one and a half from east to west.

In its fauna, physical structure, and history, it is in close agreement with the inner Kudamuttu region, and was fished in conjunction with the Kudamuttu Pars in 1828. Spat-falls have several times been recorded since 1861, namely in 1878, 1881, 1895 and 1897 when young oysters lay thick on all the rocky outcrops and wherever there was any cultch, quantities being found adhering even to the valves of *Pinna*, which are fairly abundant on the edge of the sandy ground on the western margin

The generation of oysters seen for the first time in 1897 were reported healthy and still plentiful in the following year, but in 1899 the bank was described as almost bare of oysters. A very large number of byssal cables was noticed at this 1899 inspection, indicating probably a recent inroad by rays (*Rhinoptera* sp) upon what must have been a promising bed of oysters

The Inspector, I observe, remarks that the presence of these byssal strands "shows plainly that the oysters of last year have migrated," a deduction not warranted by an intimate knowledge of the habits of the pearl oyster

Whenever an occurrence of this nature be met with, care should be taken to ascertain the condition of the individual byssal cable; we require to know whether the majority show signs of having been broken with violence as happens normally when oysters are torn away from their attachment, or if the strands of each cable join together at the free end in a pale coloured semi-gelatinous "root." Only if such "roots" be present can we infer voluntary migration, for when an oyster decides to shift its quarters it sloughs the root of the byssus; it never severs it— indeed such is an impossibility. In any case a pearl oyster's migration is hardly worthy of such a designation; at the most its journey can be measured in yards and for practical purposes the power may be ignored—a power of little advantage to the possessor except to shift position from one side of a fragment of rock to another. Thus I have seen an oyster three years old crawl our inches up the side of a stone to get away from an eddy of sand playing round the base.

11. KARUWAL GROUP

A series of the usual small rocky patches called pars lying seven miles east-north-east from Trichendur Pagoda. The depth is $7\frac{1}{2}$ to 8 fathoms.

The principal banks are Velangu Karuwal and the Karai Karuwal occupying the southern portion of the group, with the Periya Malai Piditta and Naduvu Malai Piditta Pārs on the north, the whole scattered over an area about three miles long by from one to two miles broad in an east to west direction

The rocky areas have the same general features as the other pārs of the Central division lying in a similar depth—flat-surfaced rock outcropping in patches of different size from a surrounding waste of sand

The rock is the usual somewhat variable calcareous calcrete The sand to the west of the group is fine grained and passes gradually into a chank bed. On the pār region proper the composition of the sand varies considerably, on the surface of the rock foraminifera (*Orbitolites* and allied forms) form a notable proportion of the bulk; elsewhere the grain becomes frequently coarse and occasionally grades into a distinct gravel On the northern section a considerable amount of small Lithothamnion balls is locally abundant

Among the characteristic organisms we have *Siphonochalina communis, Spongionella nigra, Axinella tubulata, Axinella donnani, Clathria indica*

A few corals, chiefly *Favia* sp (no Madrepores were seen); *Eunice tubifex* is abundant.

Other common organisms are *Pentaceros lincki, Linckia lævigata, Antedon* spp, *Ophiuroids; Scrupocellaria* sp, *Padina commersoni, Codium tomentosum*

A considerable number of dead oyster shells were found of a size of those from two and a half to three years old Living oysters of about the same age were present here and there, the majority enveloped in the encrusting mass of *Clathria indica*.

The Karuwal group has brought oysters to maturity more frequently than any other bank save the Tolayiram Pār during the last century—in 1805, 1815 and 1862 Since the last named date young oysters have appeared here in quantity at least five times, —in 1863, 1874, 1878, 1884 and 1897; no inspection of the Karai Karuwal was made in 1865, 1870, 1873, 1874, 1875, 1877, 1887, 1889–1890, 1892 1893, 1900 and 1902

The oysters found in 1897 were still on the pārs in 1899 and would have been ready to fish the following year when however the bank was not examined, owing presumably to the fishery then in progress on the neighbouring Teradi Puli Piditta Pār.

The fishery of 1862 on these banks produced a net profit to Government of Rs 1,10, 619.

The general characteristics of the Karuwal group are the most favourable of any seen during the investigation, the ground approximating most nearly to the condition found on certain of the better parts of the Cheval Par—the most valuable and reliable of all the Ceylon banks.

In both cases we find the depth of water about the same, while the bottom on the Karuwal group has a diversity in physical characters somewhat approaching that found on the Cheval, stretches of rock much broken up by patches of sand overlaid in places with a considerable quantity of cultch consisting of loose blocks of calcrete, nodular masses of Lithothamnion ("kotteipakku") and worn fragments of dead coral ("challai") Such diversity seems a condition specially suited to the requirements of oysters

12. ODAKARAI PĀR.

A bank lying six miles west of Trichendur and due south of the Karuwal group with which it appears to be linked in its main characteristics.

Much of the bottom is well cultched with Lithothamnion nodules* and the extent of rocky bottom is satisfactory, the pār extending about $1\frac{1}{2}$ mile north and south. The depth is 8 to $8\frac{1}{2}$ fathoms.

Prior to 1883, this, in common with the banks included under the term Manappad group, received insufficient attention and *there can be little doubt that fishable oysters occupied the bank in* 1900 *and perhaps in* 1901,—years when no examination was made, although it was reported in 1899 that oysters of $2\frac{1}{2}$ inches in depth were sufficiently numerous to give 20 to a dive

In the 44 years since 1860 the bank was examined sixteen times only, *so that no inspection was made during* 28 *years. Twice there was no examination for five years in succession, and this in view of the bank being for all practical purposes a portion of the most prolific oyster-maturing ground on this coast.*

---

* Inspection Report, 1887

13 CHODI PĀR

A bank four miles west of Trichendur in $8\frac{1}{4}$ to 9 fathoms of water   I had no opportunity to examine it

According to the inspection records it bore oysters of one, two and three years of age in 1869 and is described as being covered with shells and coarse sand about six inches to a foot deep in 1891 and 1894   It is marked as "useless" in the summary of 1899, a conclusion I do not think is justifiable in view of (*a*) the oysters met with here in 1869 and (*b*) its proximity (one mile north) to the Tundu Pār which yielded oysters at the fishery of 1900   It is note-worthy in this connexion to observe that these Tundu Pār oysters were not known to the Inspector prior to the fishery in question, being discovered accidentally by the divers on their way to the fishery ground on the Teradi Puli Piditta Pār.   Once again I feel driven to the conclusion that inspection work has too frequently been performed in perfunctory manner, with want of method and over too limited an area.   Only ten times since 1860 has any atten-tion been paid to this bank and in view of the imperfect method of inspection employed I am far from being convinced that the examination was efficiently carried out and that the results shown are reliable.   In most years no note is supplied of the number of dives made, and in the absence of this we have no guide to the thoroughness of the work done.   I shall return to a consideration of this vitally important subject when dealing with general conclu-sions.

14 TUNDU PĀR.

A bank lying one mile south of Chodi Pār at the same distance from land, depth from 9 to $9\frac{1}{2}$ fathoms

It appears to have been fourteen times examined in the course of the last 44 years   In 1897 it was not examined ; in 1898 oysters were "plentiful, 35 to a dive, two inches in size and healthy in appearance", the succeeding year states "Nothing of value, " while in 1900 the fishing fleet stumbled by chance on a fine bed of oysters, fully four years old on this very bank, a telling impeach-ment of the accuracy of the general results of the examination carried out in the preceding year !   The oysters plentiful in 1898, and missed at the regular inspection of 1899, would assuredly have matured and died unknown had the accidental rediscovery of the bed not been made by the divers on their way to the "official" fishing ground

The fishery of 1900 proves the good potentialities of this bank which deserves regular and careful attention in common with all the groups in this neighbourhood. It is also to be noted to the credit of this bank that the oysters fished here in 1900 were larger shells than those from the Teradi Puli Piditta Par and fetched better prices than the latter.

It was remarked that the Tundu Par oysters were covered with weed, whereas those from the other par were practically clean.*

15. MANAPPAD GROUP

Under this name I propose to include a one-ranked series of pars extending over 6 miles north-east and south-west parallel with the coast between Trichendur Pagoda and Manappad point. They lie at an average distance of 8 miles from land. The depth ranges within close limits from 8 to 9 fathoms.

From north to south the names of the constituent banks read— Trichendur Puntottam Par, Sandamacoil Piditta Par, Teradi Puli Piditta Par, Semman Patt Par, and Manappad Par, together with a few smaller rocky patches.

Prior to 1885 these banks received scant attention and were seldom examined, under the impression, I believe, that they were of little or no value. However in 1897, oysters ranging from ⅜ inch to 1½ inch in depth were found on Sandamacoil Piditta, Teradi Puli Piditta, Semman Patt Par, Surukku Onbadu Par (Manappad Par appears omitted from every inspection since 1860!) and in 1899 well-grown healthy oysters were found plentiful on all the four banks.

The following year the Teradi Puli Piditta Par was fished together with the Tundu Par already described. Unfortunately the quality of the oysters from the former par was too poor to encourage the divers to attend in large numbers and continue for a prolonged period. It appears possible that they were fished a year too early, though this is a point that was not definitely settled.

The valuation of a sample of these oysters in the October preceding was reported to be Rs. 10-2-0 per 1,000 and according to the experience of many fisheries on the Ceylon side, the actual price obtained at the fishery following is invariably considerably

---

* " Proceedings, Board of Revenue, Madras," No. 208, October 1900.

higher   In the present case Government had the utmost difficulty
in obtaining the valuation figure and indeed were we to exclude
the larger and finer Tundu Pār oysters, the price at which the
Teradi Puli Piditta oysters were sold would be found to be below
the sample valuation

It would be found of great assistance to Government and to
buyers alike if a second valuation sample of oysters were drawn
immediately prior to the fishery, say ten days preceding, in addition
to the one obtained in the October or November of the preceding
year. This is regularly done at the Ceylon fisheries and serves as
an efficient check and corroboration both of the accuracy of the
preliminary valuation and of the identity of the ground selected
for fishing with that from which the first sample was taken. As
showing the possibility of error in localization of patches of oysters
when the organization is imperfect, are the two well-known
instances of this given by Sir William Twynam, namely—

(a) How in 1836 two beds of young oysters were fished in
error instead of one bearing old and properly matured ones, and
(b) how the fishery of 1860 on the Moderagam was all but lost, a
long continued search of three days being necessitated ere the bed
was rediscovered.*

An omission which I cannot understand is the fact that no
inspection was made of the Semman Patt and Surukku Onpatu
Pārs in 1901 as they bore oysters in 1899 of the same age and in
the same abundance as those on the Teradi Puli Piditta Pār   No
examination of these was made in 1900 and it is quite probable
that patches of fine quality and large sized oysters might have
furnished a fishery on these pārs in the year named. This region
in 1901 was by far the most important to examine and for some
reason or lack of system the obvious was not carried out

SOUTHERN OR COMORIN DIVISION

Of the banks forming this division and stretching from
Manappad southwards to Cape Comorin little is known   A list of
some of these banks is given on pages 61–62   Of these only the
Manappad Periya Pār appears to have received any attention
This bank lying in 5¾ to 7 fathoms is nearly 10 miles in length by
about one mile in breadth   It lies from 6 to 10 miles off the coast,
south-east of Manappad and about 5 miles south-west of the southern

* "Report on the Ceylon Pearl Fisheries," 1902, page 20

15

extremity of the Manappad group of pārs. No information is given in the inspection summary of the character of the bottom. It would be advisable if the Inspector be instructed to pay special attention to this group during the next few inspections in order to obtain data for comparison of these banks with the better-known ones of the Central division

Historical evidence as already quoted points to some at least of these banks being occasionally productive. I know of no physical reason why such conditions should cease

NORTHERN OR KILAKARAI DIVISION.

The limits of the banks comprized in this category lie between Vaippar on the south and the island of Rāmēsvaram on the north, a distance of 60 miles. In the past considerable attention has been devoted to their examination, very much more indeed than that given to those of the Southern division which are more deserving of such care.

All these northern pārs suffer from the excessive turbidity of the sea which prevails during stormy weather. The proportion of mud present in their sand is much greater than in the case of either the Central or the Southern division, and as a consequence pearl oysters exist in a condition of chronic starvation, are stunted from an early period and never survive to a fishable age, if we may judge by the records of the past 100 years and from the effects I have noticed in those experiments where I have kept oysters under circumstances simulating a like condition of silt-laden water.

Much of this mud is derived from the rivers entering the sea between Vaippar and Pāmban, mud which moves north-east up the coast during the south-west monsoon period. In several places eddies caused by the deflection of the current by the presence of the chain of islands lying parallel with this part of the coast conduce to the formation of mud deposits at definite localities, one of which we found between Nallatanni Tivu and Upputanni Tivu Pārs; other mud deposits are marked on the Admiralty chart.

Between Nallatanni Tivu and Pāmban the banks have all the characteristics of the useless Ceylon banks immediately south of Mannar island and are distinguished by an inordinate luxuriance in growth and variety of Algae, such as *Laurencia*, *Polysiphonia*, *Corallina*, *Chrysymenia uvaria*, *Halimeda tuna*, and *Kallymenia perforata*

Such pars are, I fear, uniformly valueless and unworthy of inspection oftener than once in four years.

Greater attention is required in the southern portion of the division where there exists the possibility, rendered somewhat definite by the presence of the remains of a fishery camp on Nallatanni Tivu, of oysters some day maturing. The prospect is not hopeful but is sufficient to justify an inspection in alternate years. The pars requiring the most attention are the Upputanni Tivu, the Nallatanni Tivu and the Vembar Periya Pars. The two first lie four miles off the coast south and south-east of Valinukam Point, the last south-east of Vembar village.

The rocky bottom on all these pars is the usual brownish dense limestone calcrete, while the sand is in most cases rather finer than that from the Central and Southern divisions and the amount of mud mingled with it is very markedly greater in quantity.

When inspection of this ground be made, diving and dredging traverses should be made over the whole of the ground at depths between $7\frac{1}{2}$ and 10 fathoms to the south and east of the two Tanni Tivu and Vembar Pars. Some of the ground we met here was distinctly promising, and being further from land and at greater depth the bottom is more free from mud than on the inshore banks.

The characteristic organisms of the Tanni Tivu Pars are sponges in great abundance (see page 83 for names), various Gorgonoids, notably *Juncella juncea*, an occasional Astraeid, the tubes of *Eunice tubifex* and numbers of *Pinna, Modiola barbata* (suran) generally absent.

Kumulam Par is valueless, as are also Valinukam and Valinukam Tundu Pars and some others in shallow water between Valinukam and Vembar.

VIII.—THE ANATOMY AND EARLY LIFE-HISTORY OF THE INDIAN PEARL OYSTER

The elucidation of the main features in the anatomy of the pearl oyster was the first work I did in the Ceylon pearl fishery investigation as assistant to Prof. Sir W. A. Herdman, F R S. The results are given in detail in Volume II of the Ceylon Pearl Fishery Reports published by the Royal Society, 1903–1906, and I propose to give here merely those essential facts that are necessary to the mental equipment of the executive officer in charge of pearl bank inspection if he be not a trained biologist.

The Indian pearl oyster, *Margaritifera vulgaris*, is a small and widely distributed species belonging to the large family Aviculidæ, which is closely related on the one hand to the Scallops (Pectinidæ) and on the other to the mussels (Mytilidæ). All these exhibit a number of primitive characteristics whereof the lining of the shell in the case of the aviculids and mytilids with a lustrous pearly layer—nacre—is one.

Our local pearl oyster seldom exceeds 3½ inches in height and slightly less than this in length. It is a typical bivalve mollusc, and might with advantage be made a teaching type, so simple in form and easily distinguished are the various organs. The shell consists of a right and a left valve attached together along one side, the hinge, by an elongated dark brown elastic pad, the ligament. The hinge line is shallow at first, but with increasing age becomes deep and gutter-like ; its depth and width are our best indications

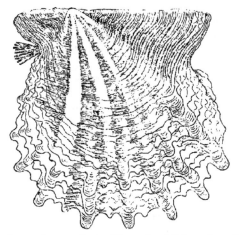

FIG. 1.   Indian pearl-oyster, *Margaritifera vulgaris* Schumacher   Natural size.

of the age of the oyster. Each valve is roughly circular (sub-quadrangular) with a tendency in the old to become slightly oblong ; an ear-like projection (auricle) is given off at each end of the hinge (fig. 1). The two valves are somewhat asymmetric, the left valve being deeper and more convex ; the margin of the right valve is the more flexible, so that when the shell is tightly closed, the right valve is drawn partly within the left, the margins of the two valves being brought together to a width of nearly half an inch all round. This gives additional protection against unwelcome intruders. When young and in vigorous growth the ventral margin of the valves is prolonged into finger-like projections set at fairly regular intervals (fig. 2). Their growth is discontinuous, so that as

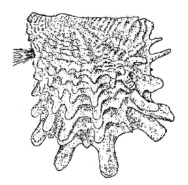 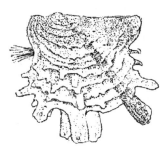

Fig. 2. Two young pearl-oysters showing free and vigorous growth and a great development of the marginal digitations. In the second figure a strong ray of dark colour passes obliquely across the shell from the umbo. Natural size.

the later ones arise and project further beyond the earlier, we get eventually a series of radial lines of overlapping digitations running from the umbo—the prominent region forming the beak of the shell—outwards and downwards to the lower or ventral edge. The older gradually get worn away by attrition and in oysters over three years old, especially when crowded and suffering from shortage of food, renewal becomes slow and may even cease, eventuating in a smooth-surfaced, even-edged condition indicative of senility. In such oysters, the hinge groove is deep and wide and the "ears" almost obliterated. Coral and sponge growth upon the valves is a frequent feature and as is natural, many are attacked by the burrowing sponge *Cliona*, which so riddles them with intricate tunnelling as to impart the appearance of being "worm-eaten," the technical name given such shell by the dealers in mother-of-pearl. Five years appear to be the ordinary span of life of the Indian oyster; the records of inspection show that if oysters be not fished within this age, the bed may disappear by the following season. This comparatively short life involves a corresponding small size in the pearls produced, the larger oysters of Australia—the Gold lip (*M. maxima*) particularly—have a far longer life and in consequence the maximum and average sizes of pearls produced by them are much greater, but the lustre and skin are not so perfect as in the Indian and Ceylon pearls.

The shell substance consists of three principal superimposed layers, with a fourth at those places where the muscles of the body are attached to the inner surface of the valves. The outermost layer, the periostracum, is tough, thin and horny, and yellowish in colour, the second, the prismatic layer, is thicker and consists of tiny columnar prisms of carbonate of lime, arranged at right angles to the surface of the shell; in this layer are the red and yellow pigments that give bright colouring to very young shells and a reddish mottling to the older ones. Both these layers are secreted by the cells on or towards the edge of the mantle as will be described later. The nacreous or third layer is the one that yields mother-of-pearl. Its lustrous, white, iridescent appearance is characteristic, and is due to the way in which its substance is built up of tiny and slightly irregular lamellæ deposited horizontally and overlapping one another. As a result the surface shows a fine lineation that causes a deflection of light rays, resulting in the peculiar lustrous iridescence associated with this substance. True pearls consist of the same material. Both the periostracum and the prismatic layer are limited in their growth as they are produced *normally* only by the cells near the mantle edge, nacre on the contrary being secreted by the greater part of the exterior surface of the mollusc's body, increases continuously throughout life, at first both superficially and in thickness, and later, when full size has been attained, in thickness only. Thus in old oysters the nacreous layer may attain quite a considerable thickness, particularly in the inner region near the hinge. The fourth layer, or hypostracum, is closely akin to the prismatic layer in structure, being built up of distinct columnar prisms; it covers only those portions of the shell to which muscles are attached. The largest area of hypostracum is that beneath the insertion of the great adductor muscle, towards the hinder central region of each valve. It has much importance in the origination of muscle pearls, which comprise the greater number of seed and baroque pearls as we shall see below.

If we watch a living pearl oyster in an aquarium tank, we see that in the normal condition it lies with the valves slightly agape, their edges fringed with a feathery frilling along each side of the opening. The oyster seldom opens more than ⅜th of an inch ; but this is ample to permit of breathing and feeding Like the common sea mussel, the pearl oyster is attached throughout life to some solid object , when much crowded they adhere to one another in clusters.

The anatomy is easily understood if we remove one valve, say the right, together with one gill and the thin lining of tissue adhering to the inner surface of the valve This delicate layer is one-half of the mantle or pallium and is in reality a simple fold of the body wall of the animal. With its companion fold on the other side, it envelops the whole visceral mass of the animal much as the flaps of a coat enwrap the body of the wearer. The external surface of each mantle fold is attached lightly to the inner surface of the shell, and is equal to it in extent when in the uninjured and unretracted condition. Its edge is thickened and reflected inwards into a second edge. The cells of the true edge secrete the horny periostracum, those of the reflected one are mainly tactile, the most sensitive being borne on delicate dendritic processes that lightly interlock with those of the opposing mantle edge, thus forming a most sensitive *cheveux-de-frise* guarding the entrance to the cavity enclosed between the mantle folds—the pallial cavity.

The mantle itself consists of a thin matrix of loose connective tissue containing blood spaces, muscle fibres arranged fan-wise for the retraction of the mantle edge, and a fine network of nerve fibres. Externally where it is in contact with the inner surface of the shell, it is covered with a layer of brick-shaped secreting cells, the external pallial epithelium These normally secrete the mixture of carbonate of lime and shell gelatin (conchyolin) which crystallizes into nacre after it is excreted. But it is notable that this same epithelium that normally secretes nacre, possesses the power to secrete all the other layers if need demands This I proved experimentally both by injuring the edge of the shell and by removing pieces from the centre of the valve, exposing the ectoderm. In both cases, a layer of periostracum was quickly formed ; only after an interval of several days was the secretion of nacre resumed.

Turning now to the mass of the oyster lying upon the left valve from which the right valve, gill and mantle fold have been

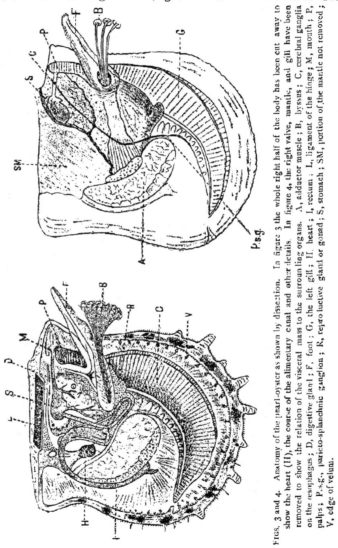

FIGS. 3 and 4. Anatomy of the pearl-oyster as shown by dissection. In figure 3 the whole right half of the body has been cut away to show the heart (H), the course of the alimentary canal and other details. In figure 4, the right valve, mantle, and gill have been removed to show the relation of the visceral mass to the surrounding organs. A, adductor muscle; B, byssus; C, cerebral ganglia on the œsophagus; D, digestive gland; F, foot; G, the left gill; H, heart; I, rectum; L, ligament of the hinge; M, mouth; P, palps; P.s.g., parieto-splanchnic ganglion; R, reproductive gland or gonad; S, stomach; SM, portion of the mantle not removed; V, edge of velum.

removed, we have the appearance I have drawn in figure 4.
Occupying the hinder portion of the centre, is the great crescentic
fleshy mass of the adductor muscle, which passes athwart the body
from valve to valve and by its contraction closes the shell. Beneath
and to the front is a conspicuous striated sickle-shaped organ, the
gills or branchiæ Above the adductor and the gills is the swollen
yellowish visceral mass—the body proper In it are contained
the greater part of the alimentary canal and of the nervous system
together with the main glands, those connected with digestion (the
so-called liver) and with reproduction Behind this mass is a
triangular cavity—the pericardium--bounded below and behind
by the adductor muscle , in it is lodged the heart, made up of two
dark pouch-like auricles below, and one yellowish strong muscular
ventricle above ; its upper wall enwraps for support a section of
the intestine, which thus has the appearance of passing through
the heart From the anterior side of the visceral mass a tongue-
shaped organ, the foot, protrudes, passing beyond the shell, when
extended, through a small gap, the byssal sinus, between the two
valves just under the anterior " ear " of the shell The foot is of
great importance to the pearl oyster ; by its aid it can drag itself
slowly along when it desires to shift position and also can attach
itself securely to some stable foothold The foot is highly cavernous, full of spaces that can be injected and inflated with blood
at will Along the lower surface is a groove ending behind in a
deep pit If the oyster has detached itself from its previous
foothold and desires to crawl, the foot is injected with blood and
the tip pushed forward to its full extent, while the sides of the
groove at the fore end flatten and produce an adhesive surface.
On this purchase, the animal drags itself forward and then while
retaining a hold at the hinder end of the foot releases the tip and
pushes forward another quarter of an inch Progress is slow and
any continuous journey seldom exceeds a few inches When the
animal desires to refix itself, the foot is again employed for the
purpose. In this case, the sides of the groove come together and
form a temporary tube, except at the fore end where a small space
remains open upon the surface to which attachment is to be made.
The cells lining the groove now pour forth a gluey secretion which
sets firm and elastic and takes the form of the tube within which
it was produced. At the end of about five minutes, the byssal
groove opens and reveals a stout golden green fibre ending at

16

one end in a circular disc of attachment upon the solid object to which the oyster is affixing and at the other in the deep byssal pit sunk in the base of the foot The process is repeated indefinitely till a strong cable of byssal threads—the byssus—has been formed. I have watched young pearl oysters repeatedly form within an hour or two, a tuft of byssal fibres and then free themselves for another journey by casting off the root from the byssal pit.   Old oysters do not discard their byssus and form new ones save under stress of circumstances.

The *alimentary canal* is comparatively simple in form and structure   The mouth, a simple opening guarded on either side by two long flaps which embrace at their lower ends the tips of the gills, opens into a straight horizontal gullet which in turn leads into the stomach, located in the centre of the visceral mass.   The digestive gland, dark green in the living condition, surrounds the stomach and empties its secretion thereinto by several ducts. The anterior section of the intestine forms a loop in the lower part of the visceral mass and then merges into a wide posterior section which passes through the ventricle and then turns downwards as the rectum along the hinder surface of the adductor muscle, to end in an anal orifice prolonged dorsally into an ear-shaped directive projection.   Parallel with the straight part of the anterior section of the intestine is a long diverticulum, lodging a long crystalline style, whereof the outer end projects into the stomach.

The gills are two in number, each folded upon itself longitudinally in such a manner that there appears to be a pair on each side   Each is made up of a multitude of slender tubular filaments placed side by side.   No cross bars or branches connect them. When dead, the filaments are readily separated, but in life opposing pads of strongly ciliated cells on adjacent filaments form an interlocking mechanism that prevents displacement, an arrangement somewhat analogous to that seen in the vane of a feather. Respiration is effected by a current of sea-water, admitted through the open margins of the shell and mantle, and set up by the rhythmic lashing of cilia covering the inner surface of the mantle folds and also upon the gill filaments   By this means a continuous current of fresh sea-water passes between the gill filaments into the suprabranchial cavity and thence outwards, past the anus back into the sea   In its passage through the gills the water in bathing the tubular filaments containing blood, parts with some of its oxygen and receives waste carbon dioxide, so effecting respiration. The general inner surface of the mantle also participates in this function, for it too is highly vascular and separated from the incurrent water by an extremely tenuous epithelium

The pearl oyster feeds upon the minuter organisms found in the plankton of the surrounding water There is at least partial selection exercised, chiefly by means of the filaments along the mantle edge, for these bar the way to copepods and small worms and similar small animals that might be of more trouble than use. But all algal matter is admitted freely. Whatever organisms pass in are carried with the current into the natural filter formed by the gill filaments and there are collected by certain sets of cilia and directed into definite ciliated pathways leading towards the palp and the mouth ; these progress much like a football trundled along the ground When the particles reach the palps they are rolled together to form a tiny pellet and this is finally forwarded by a ciliated pathway into the mouth

The excretory system is a pair of simple tubular chambers lined with secretory cells Each opens at one end into the pericardium, at the other to the outside.

The blood system—In the pearl oyster the blood is colourless It passes at each pulsation of the ventricle into two main channels or arteries, one leading forwards, the other posteriorly. These arteries distribute blood to all parts of the body ; thence it passes to the gills and the mantle to be purified and from there is returned to the auricles. Only purified blood passes through the heart

As pearl oysters are sedentary animals, mere passive machines for eating and reproducing, and are well protected against all but particular enemies by the strength of their shell, they require no elaborate nervous system As a consequence there are no conspicuous sensory organs and in particular no eyes In this last respect they are less specialized than their near relatives, the Scallops, for these have very numerous well-developed eyes upon the margins of the mantle Three nerve centres exist, each composed of a pair of ganglia (small rounded aggregations of nerve cells). One is found on the sides of the oesophagus, one at the base of the foot and the third on the anterior face of the adductor muscle. These centres are joined by long nerve trunks, the connectives, and from them arise a network of delicate nerve fibres running to all parts of the body, but chiefly to the muscles and the mantle edge, for active co-operation has to be maintained by these organs; if the mantle filaments sense danger, they signal to the ganglia and these transmit an impulse to those muscles to be actuated in defence—usually the retraction of the mantle edge, and the contraction of the adductor whereby the shell is closed

The sexes are separate, but no outward differences are discernible in the shells of the two sexes The gonads or reproductive glands consist of myriads of glandular tubules enveloping the digestive gland and the alimentary canal. In the males the gonads when ripe give a pale yellow tint to the surface of the visceral mass ; in the female the colour is usually deeper, inclining to orange As the products ripen they pass into larger ducts and these in turn empty into two main tubes communicating with the exterior, one on each side near the anterior ends of the gills.

Ova and spermatozoa are poured out by each sex haphazard into the sea, where the ova have to chance fertilization Probably there is not such waste as one might infer, for these oysters form beds counted by the million and the emission of either product acts, as I have actually observed, as a direct stimulus to those of the opposite sex that have not as yet emitted April May and September October are the maximal spawning seasons, but like the edible oyster (*Ostrea virginiana*) of Indian back-waters, some individuals can usually be found in a ripe condition at almost any time.

The ova when emitted are flask-shaped, the neck forming the micropyle. After a spermatozoan has forced entrance into an ovum through the micropyle, thereby effecting fertilization, the latter becomes rounded, and soon begins to segment. Segmentation is unequal A trochophore form similar to that familiar to us in many allied molluscs, as well as among the polychaet worms, constitutes the first larval stage , first appears a zonal and pre-oral band of short cilia, followed a little later by the development of a long apical tuft and a small posterior patch. By the 20th hour this larva is fully formed and already shows a short enteron By the third day at latest, probably somewhat earlier when conditions are favourable (the present results are based upon observations which I made on larvæ living under artificial conditions in glass bowls), the second larval stage, the veliger, is attained In this a minute ovoid embryonic shell has appeared, colouriess, transparent and

structureless. From beyond the widely opened edges of this
minute bivalve shell protrudes a large lobe, the velum, fringed

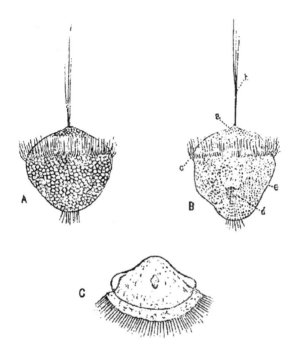

FIG. 5. Larvæ of the Indian pearl-oyster. A. and B. trochophore stages ;
C. earliest form of the veliger stage. Very highly magnified.

with cilia, a modification of the pre-oral region of the trochophore
larva. Soon various organs begin to make their appearance as
shown in figure 6. In these larvæ, as yet free-swimming, the need
for sense organs is greater than in the sedentary adult ; so we find
developed a pigmented spot, a rude eye, and an otocyst composed
of a tiny sac containing minute refractive granules. Developing
gill filaments appear, together with a bilobed digestive gland, and,
finally two adductor muscles, one at each end of the body.
Advanced larvæ at this stage constitute the "spat" that replenishes
our pearl banks by means of a "spat-fall." The young oyster is
now ready to settle down upon the bottom and begin its sedentary

existence. This may happen as early as the fifth day after fertilization, but may be delayed considerably if conditions for the
change be unfavourable, as for example the prevalence of a strong
current or stormy weather.   After settlement, the velum is retracted
and eventually absorbed and growth of the embryonic shell ceases.
In the course of a day or two, shell-growth of a different character,
marked out into a minute hexagonal pattern, is observable around
the margin of the valves; this is the first appearance of the

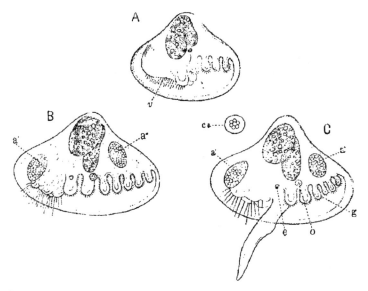

FIG. 6. Later phases of the veliger stage. a', anterior adductor muscle ; a'', posterior
adductor ; e, eye-spot ; o, otocyst ; g, embryonic gills ; v, Velum; c', enlarged
view of otocyst containing granules.

prismatic shell-layer. Eventually growth produces such a form as
is seen in figure 7, B. The spat is now about one millimetre
across, and three regions are distinctly visible in the shell,
(a) the clear embryonic shell marked by delicate lines of growth,
and forming the hinge, (armed it should be noted with a number of
minute teeth) and the prominent umbo ; (b) the now very extensive
prismatic region extending to the margin all round and gradually
producing the adult form of the shell, and (c) an internal and

intermediate region lined by nacre. Eventually the embryonic shell
becomes imbedded in the adult shell and ceases to be distinguishable. At an early stage the byssal sinus appears as a tiny bay at
the anterior side of the shell. Even in the free-swimming stage the
foot had appeared and as soon as the spat settles, this organ and its

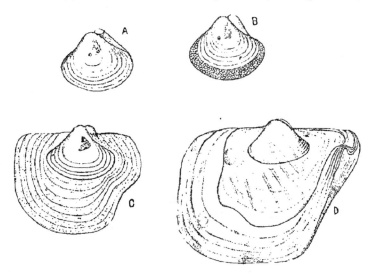

FIG. 7. Pearl oyster spat in progressive stages. A. form at time of attachment ;
    B. shows the beginning of the prismatic layer ; C. and D. very young forms
    of the immature oyster, the byssal sinus already distinguishable. In D.
    the gill filaments are visible.

byssal gland begin to function by producing tiny byssal threads to
serve as attachment cables. For some days after settling the spat is
actively locomotive, continually throwing off its byssus and mov-
ing to a new foothold. Little by little this activity slackens but
under uncomfortable conditions even the adult may repeatedly cast
off and renew its byssus, crawling slowly to a new position in the
interval. I have seen this happen up to eight times in fourteen
days, in the case of a marked oyster kept under observation in an
aquarium tank.

It is of great practical importance to be able to distinguish the
true pearl oyster spat from that of its near relative, *Avicula vexillum*,
called most appropriately "False-spat" in our records. Time

after time it seems that wide deposits of the spat of this mollusc, clustering on weeds and stones, have been mistaken by former Superintendents of the Pearl Banks for the true pearl oyster. For reference, I give an enlarged drawing of a young *Avicula vexillum* at the age when it has been so mistaken (Fig. 8 B) Figure 8 B of the form of true pearl oyster spat is provided for comparison.

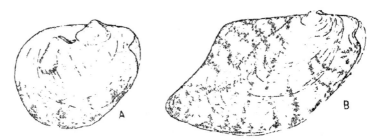

FIG. 8 Comparison of the spat of the true pearl oyster (A), with that of *Avicula vexillum*, the so called " false spat " (B).

Examination shows differences sufficiently distinct to render descrimination fairly easy The chief of these are in the colouration and the outline. False spat is more brightly and profusely coloured with reddish radiating bands, and these occur from an earlier age More distinctive still is the markedly oblique form of the false-spat, and particularly the way the lower posterior angle is prolonged backwards into a long rounded point. In pearl oyster spat of the same age, the colouration is distinctly less abundant and clearly marked, while the obliquity of the valves is slight As in the adult the tendency is towards a rounded subquadrangular outline These differences however do not show themselves clearly till the spat attains a length of about two millimetres; in the veliger and very early fixed stages there are no distinguishable differences between the larvæ of the two species

---

IX—PEARL FORMATION IN THE INDIAN PEARL OYSTER.

In the following brief note, I do not propose to traverse the past history of the theories and researches upon this subject; the conclusions to which I have come after twenty years' work on this

and related problems are sufficiently definite and, I believe,
conclusive, in the light of the most recent results obtained by
others, to stand by themselves as strictly conformable to the actual
facts.

Pearls in the widest sense may be defined as more or less
rounded masses of shell substance made up of concentric layers
laid down around a nucleus The shell substance may be of any
one of the four layers normally present in such shells as the pearl
oyster or two or more of these may alternate in the layers Some
pearls consist wholly of periostracum ; these are brown and on
account of the lack of lime in their composition, they frequently
crack as they yield up moisture In one example of a periostracal
pearl in my possession, half the sphere is coated with nacre ; had
the process been extended and continued, a complete coating of
nacre would have been deposited, converting a valueless pearl
into one of considerable price, but of specific gravity less than
normal Periostracal pearls are formed invariably in or close to
the edge of the mantle, where are situated the cells normally
engaged in the secretion of periostracum Nacreous pearls
characterize the pearl oyster, but in molluscs where the inmost
layer is porcellanous, pearls produced are themselves porcellanous ;
examples of these are the well-known pink pearls obtained from
the West Indian conch, *Strombus gigas*, the rare and beautifully
watered pearls produced by the chank (*Turbinella pirum*) in our
own waters, and also the lustreless white pearls sometimes found
in the edible oyster Hypostracal pearls are, in my experience
the most numerous of all in the local pearl oyster, but they are
usually minute and even microscopic They were called calcospherules in the Ceylon Pearl Reports They occur, when present,
in and around the insertion ends of the pallial and adductor
muscles, often in great abundance ; "nests" of 20 to 50 are not rare
when properly sought for Many of these become the pseudo
nuclei of nacreous seed-pearls, the real nuclei being of course the
nuclei of the calcospherules themselves Not infrequently conti-
guous pearls of this nature fuse into a compound mass of irregular
shape, one form of the baroque pearl, useful to the imaginative
jeweller for the production of quaint pearl ornaments. One such
compound mass I have seen worked into the form of a mulberry
fruit, mounted with a spray of golden leaves Other artists have

17

utilized such masses in the production of grotesque figures when from time to time jewellery of this design is fashionable

True gem-pearls are those composed of lustrous nacre and of symmetric shape, round or pear-shaped preferably These are produced normally in the mantle in the region between the pallial line (the curved line marked by a row of muscle scars) and the limit to which the deposit of nacre extends marginally—from half to three quarters of an inch inwards from the free edge of the shell Such pearls seldom occur in the visceral mass area of the mantle or within the muscles As will be seen later, these gem-pearls have frequently some foreign intrusive body as the nucleus, whereas the less valuable pearls found in and around the muscle insertions have some particle produced by the oyster itself, as the centre of deposition

In all cases an envelope of secreting tissue—the pearl-sac—surrounds the developing pearl In the case of gem-pearls this arises usually as an invagination of the external epithelial layer, for the intrusive foreign body is generally found in the first instance between the inner surface of the shell and the secretory surface of the mantle. The latter being delicate yields readily to the pressure of the intrusive body which then comes to lie in a pit within the mantle substance. At first this pit is wide-mouthed, but as the foreign object sinks deeper in, the mouth of the pit narrows to a neck, and eventually may close; the next stage is for the cyst containing the intrusive body to separate from its connexion with the ectoderm and to assume a saccate shape conformable with that of the enclosed body. For this reason, Prof Herdman and I named pearls of this origin 'cyst pearls' in contradistinction to the small and usually irregularly shaped 'muscle pearls' formed within the muscles This classification has the merit of simplicity and I see no reason to amend it

Cyst-pearls in number are relatively very scarce as compared with muscle pearls, and large cyst-pearls, the true gem-pearls, are again relatively much scarcer than small sizes The former constitute the so called Orient pearls, pre-eminent above all for their lustre and purity of colour and for a peculiar suggestion of translucency not seen in other pearls

The origin of these pearls has been a battlefield of theory in the past ; the resultant confusion appears to me to be due in large part to the lack of recognition that there are these two main categories of pearls, differing in origin, and that in the case of cyst-pearls the causative body may and usually does differ with the locality and the species investigated. In the case of certain mussels (*Mytilus edulis*) the causative nucleus has been found in certain beds in France to be a larval trematode worm (Jameson and Boutan), and in certain fresh-water mussels in one locality this is replaced by a little commensal mite (Küchenmeister). In the case of the Ceylon and Indian pearl oyster, Prof. Herdman and the author found it in many cases to consist of the dead body of a larval Cestode. To this we gave the name *Tetrarhynchus unionifactor*, and we correllated it with an advanced larval Tetrarhynchid of typical form found commonly encysted in the walls of the oyster's intestine. At a later date we discovered that the adult of the latter worm is found in the sexually mature condition in the intestine of an oyster-eating ray, *Rhinoptera javanica*. At one time we intercalated an inter-mediate host, one of the file-fishes (*Balistes mitis*) but eventually the species found in the file-fish was found to be of a distinct species, not parasitic in the larval condition in pearl oysters. I have however come now to the conclusion that the spherical cestode

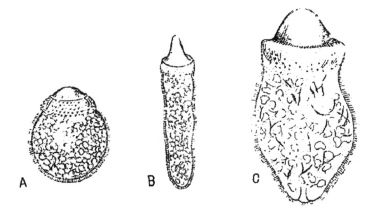

Fig. 9 Three cestode larvæ extracted from cysts found in the tissues of the pearl oyster (Gulf of Mannar). A. is the youngest stage found ; B is an elongated form (older) occasionally found, while C , seen under higher magnification and slightly compressed, shows the beginnings of a vascular system and also a terminal excretory pore. Note in all the minutely spinous nature of the collar and the multitude of tiny calcareous granules densely filling the body region.

larva (Fig 9), found in abundance in the tissues of the pearl-oyster and frequently as a nucleus in cyst-pearls from the same mollusc, is not a younger stage of the undoubted Tetrarhynchid larvæ encysted in its intestine (Fig. 10). Possibly it is the larva of a species belonging to the cestode genus *Tylocephalum* but this is a subject for further investigation.

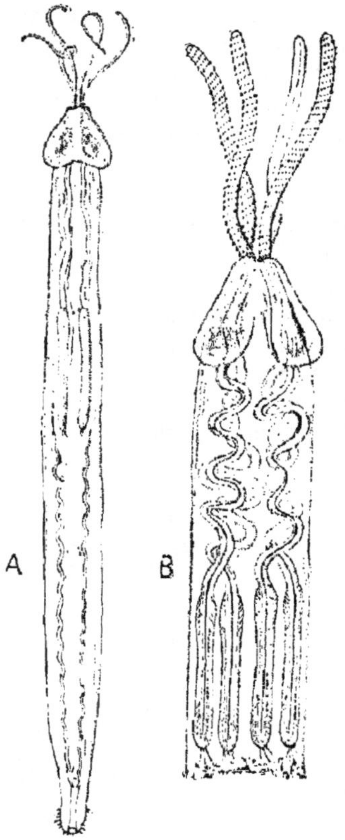

Few pearl-oysters are free from this parasite. Usually the gills contain hundreds, often very minute and never differing in any appreciable degree from those shown in Fig. 9. The digestive gland is another favourite location for these cysts, opalescent white spheres conspicuous in the dark green of the gland. In figure 11 are drawn two nuclei which I obtained by decalcification of small Orient pearls; there can be no question as to their identity with the spherical larvæ found alive in the tissues. Neither Prof. Herdman nor I ever claimed that all cyst-pearls have such nuclei; we recognized that other foreign bodies, notably grains of sand, occasionally function as the intrusive irritating factor and become pearl nuclei. We have also even found a small nematode worm coiled upon itself, forming the nucleus. So far we went, over 16 years ago. Subsequent investigation shows me that a further qualification is necessary whereby cyst-pearls may be divided into two sections, the one comprising pearls induced by the irritation of foreign bodies and the other those with

FIG. 10. *Tetrarhynchus unionifactor.* Larval form found encysted in the wall of the pearl-oyster's intestine. A. view of the entire worm. B. Enlarged view of the fore end, to show the 4 proboscides and their sacs.

A   B

nuclei of periostracal-like substance derived from the oyster's own tissues. The former class comprises, according to my investigations, the majority of the larger cyst-pearls, the latter of the smaller ones of this description which, as I have indicated above, constitute by far the larger proportion of cyst-pearls. This conclusion to

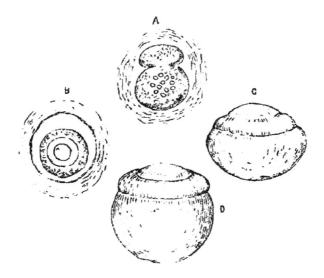

Fig. 11 A. and B. are two nuclei obtained from Ceylon Orient pearls by decalcification. In each a portion of the inmost conchyolin layers is shown In A some of the minute calcareous granules seen in Fig 9 are observable, while B. shows the collar and proboscis (face view) of this larva very distinctly C is the enlarged appearance of the pearl that enclosed nucleus B , note its resemblance in form to the outline of the spherical larva shown in D. (cf. Fig. 9-A)

our local researches disposes satisfactorily of certain objections levelled at the cestode theory and places the latter in its proper perspective ; we see that cestode larvæ, though less frequently the cause of pearl formation than was at first believed, are nevertheless the *most important factor in the production of the larger and finer of orient pearls* and therefore of supreme importance from the economic and commercial view point. Let us now see how pearl formation proceeds (*a*) in cyst pearls formed around intrusive foreign bodies, (*b*) in those with a fragment of periostracum as nucleus, and lastly (*c*) in muscle pearls.

Some of my earliest experiments made in Galle in 1902, have direct and fundamental bearing on this problem. These were in respect of the power of the oyster to repair injuries to the shell. They resulted in demonstrating that epithelial cells are capable, at least over the nacre-secreting area of an alteration in the character of their secretive power upon emergency. Thus I found that if a fragment of shell in the centre of the valve were removed, exposing the mantle which previously had been engaged in secreting nacre, the first repair substance formed was not nacre, but a yellow parchment-like material apparently identical with periostracum. Only after a stiff layer of this was formed, was there a resumption of nacre secretion. Now in all the pearls I have examined and notably in button pearls formed after the old Chinese method, and within recent years refined and extensively employed on a commercial scale by the Japanese, I have found that the nucleus, whether it be a cestode larva, a grain of sand or a spherule of mother-of-pearl (as in the Japanese culture pearls), is not overlaid directly by a nacreous layer, but has interposed between its surface and the eventual layers of nacre, a distinct and well marked deposit of stiff yellow membrane identical with repair periostracum, which indeed it is. It is evident that the intrusion of any body into the ectoderm must affect it in a similar manner to that caused by a direct injury, such as a fracture of the adjacent shell would do, hence the impulse of the cells around the intrusive body is to pour out the primary secretion employed to meet such an eventuality. The inmost layer of such a pearl is invariably of periostracum. Only after the effects of the shock have passed and normal conditions are restored, does the nacre secretion begin to be again deposited. What seems to me to be the explanation is that the membrane repair substance is really the conchyolin basis of nacre with the lime salts withheld. In other words, after a shock, the epithelial cells intermit the secretion of lime salts, but continue the secretion of conchyolin, thus giving a periostracal appearance to what would normally be a nacreous layer (conchyolin + carbonate of lime).

Another deduction which I have made from the investigation, is that only dead or dying parasites excite an irritation of the character necessary to induce pearl formation. A living parasite does not irritate the tissues in the same way, indeed it merely induces the formation of a tough connective tissue sheath or cyst enveloping it, wherein it lies quiescent and harmless, giving no further irritation. But in the case of a parasitic larva that arrives in the epithelium in a dying condition exhausted or perhaps smothered in the secreted fluid poured out by the epethelial cells a different situation is found. Instead of being within a layer of connective tissue, it lies in a depression of the epithelial cells and these act differently from connective tissue cells—with a correspondingly divergent result.

In regard to the second and more numerous class of cyst-pearls, usually however much smaller in size than those of the first class, decalcification shows no definite nucleus other than a tiny amorphous scrap of brownish refractive substance, similar apparently to periostracum Rubbel of Marburg has investigated the origin of pearls with a similar form of nucleus, obtained from freshwater mussels He showed that granules of the same appearance not infrequently appear in the secreting epithelial layer of the mantle. These at times appear to cause an irritation that induces the adjacent cells forthwith to begin the deposit of nacre upon these refractive bodies ; later, by radial division and multiplication, these cells form a minute pearl-sac around each nuclear body, which continues the deposit of concentric layers of nacre and thereby produces a pearl The same sequence of events occurs in the Indian pearl-oyster eventuating as above stated in the production of the majority of cyst-pearls found in the mantle The irritation produced is so slight that no shock is experienced and therefore no periostracal repair substance is deposited prior to the first nacreous layer.

The third class, muscle pearls, remains for consideration From their place of origin being invariably close to the insertion of muscles attached to the shell and from the columnar nature of their pseudo-nucleus, we may infer that their initial origin is due to the dislodgment of a tiny particle of hypostracum from the insertion surface of the shell, caused by some exceptional strain set up, such as an excessive and sudden contraction of the muscle involved. A particle set loose in this manner causes irritation in the same way that an intrusive foreign body or an unwanted particle of periostracum does and with similar effect : a minute pearl-sac is formed, enveloping the particle, which however in this case begins by secreting columnar hypostracum instead of periostracal membrane or nacre. This is consistent with its natural function

In the nests of these pearls in a very early stage which I have often examined, the columnar structure is extremely clearly shown ; in optical section each pearl is seen as an oval body made up of lines radiating from an almost imperceptible nucleus of the same refractive index. The pearl-sacs of adjacent pearls are very prone to coalesce both in this early stage, the calcospherule stage as Prof. Herdman and I named it originally, and at a later date when of larger size. In this way irregular pearls are formed. Usually only one concentric layer of hypostracum is laid down, but more may, rarely, be deposited. As a rule the next layers laid down are composed of nacre, and in this way the pearl finally assumes the normal appearance of a pearl, at least one of inferior quality. Owing to the crowding together of these pearls as happens normally, mutual pressure adversely effects their shape. Frequently they are found partially coalesced, as twin pearls ; more usually they are irregular and mishapen in varying degree. They constitute the bulk of the seed-pearls put upon the market.

Very vigorously grown adult oysters, particularly those of exceptional size, are prone to form nests of these muscle pearls ; this is exactly what we would expect reasoning from the facts stated above. In oysters of this description, the strength of the muscle fibres is exceptionally great and as a consequence dislodgment of tiny particles of the hypostracal layer to which they are attached and disturbance of the cells secreting this substance are more frequent than in smaller and weaker oysters, where muscular force is distinctly feebler.

X.—CONCLUSIONS AND RECOMMENDATIONS.

A.—CONCLUSIONS.

The outstanding conclusions of supreme importance to which my investigation of the records and natural characteristics of the Tinnevelly and Madura pearl banks has led are that the banks have latterly given inferior returns owing to —

(a) The imperfections of the present methods of inspection, and

(b) Deficiency in the supply of divers when fisheries are held.

(a) IMPERFECT METHODS OF INSPECTION.

The Ceylon banks have certainly enjoyed a larger measure of supervision and a more developed inspectional organization during the past half century than the Madras banks and probably during the preceding 50 years as well. They were, however, inspected in a very imperfect manner till the early sixties, when Captain Donnan introduced improved methods. Prior to that time, owing to the charts in use being imperfect and the landmarks insufficient in number and in conspicuousness, the Inspectors relied in great part upon information supplied by native headmen. The boats employed were often ill-adapted to the purpose and the search for beds was not conducted with anything approaching scientific precision.

As already mentioned, Captain Donnan, who was Inspector from 1863 to 1902, organized matters on an improved basis and so far as nautical knowledge permits brought the mechanical part of the inspection to a high level of excellence. By the preparation of large scale charts, whereon he plotted every landmark of value and the outlines of many of the pārs, he was enabled to dispense with the services of the headmen; he abolished the unhandy "ballams" which served as the inspection divers' boats, introducing in their place a handy type of whale boat; he elaborated an admirable system of "circle inspection" capable of supplying detailed information in regard to the minute features of the ground inspected—the respective numbers of old and young oysters present, the ratio of sandy ground to rocky and the distribution of oysters over it. He trained intelligent natives (Paravas) to act as coxswains of these boats and to record in diagram form the results of each and every dive made during the day's work.

The present state of the Tuticorin inspection organization is similar to that characterising Ceylon inspection prior to the inception of Captain Donnan's improvements. Charts are imperfect and do not show the position of the chief landmarks, * native pilots (pār-mandadais) have to be employed; circle inspection is not carried on in an adequate and systematic manner; native boats are still employed for the divers' use and no attempt has been made to train efficient coxswains to keep records of the work done with exactitude. As a consequence of such imperfect methods I am convinced that beds of oysters have been missed and fisheries

---

* See sketch plans I and II in annexures.

18

lost from time to time. Such mischances certainly did happen on
the Ceylon side under similar conditions. In the experiences given
by Captain Steuart in his "Account of the Pearl Fisheries" his
belief is several times stated that beds of oysters had repeatedly
been missed and that even in 1836 a bank was lost and two or
three beds of young oysters fished by mistake. He "attributed
this in great measure to the clumsy boats used for inspections and
the ignorance of the native headmen." Sir William Twynam
("Report on the pearl fishery of 1888," page 13) in commenting
on this has no doubt that want of proper landmarks, incorrect (or
rather confused) compass bearings, incorrect charts and unsatisfactory inspections had a great deal to do with such lost fisheries.
In Ceylon the classical instance is the failure of the official inspection in November 1903. At the end of the fishery in the preceding
April, plenty of oysters were present on the banks, but in
November, the Master Attendant of Colombo, the *ex-officio* Inspector, could not locate them and advised Government that no fishery
was possible in 1904. Doubts were cast on the accuracy of the
inspection and the writer was asked to direct a fresh inspection
on biological lines. Immediate success was obtained and a
splendid fishery followed in 1904, giving the Ceylon Government
a net profit of nearly ten lakhs of rupees, which otherwise would
have been lost.

Until the present day, a sea-faring education has been
considered the fitting mental equipment for Officers in charge
of the Pearl Banks of Ceylon and India. Men who had passed
their youth and early manhood on the sea were appointed,
the impression being that nautical knowledge and elementary
marine surveying were the chief qualifications for these duties.
Captain Donnan has been, without doubt, the ablest of these nautical
Inspectors, but far as he carried the improvement of inspection
methods, lack of biological knowledge prevented him from so
economizing his time as to enable him to examine each season the
whole of the potential oyster-bearing ground in his charge. In this
way it was that often enough precious days and weeks were
devoted to the examination of ground which a biologist would have
decided at once to be unworthy of detailed circle inspection, while
other large areas, biologically more favourable to oyster growth,
had to be left wholly or partially unsurveyed for want of available
time.

A concrete instance of the imperfection of present inspection methods on the Tuticorin banks is afforded by the fishery held in 1900. The bed to be fished was the Teradi Puli Piditta Pār off Trichendūr; fishing went on there for three days, but, on the fourth, some of the boats, owing to a strong head wind, were not able to fetch the proper bank and anchored three miles away on the Tundu Pār, where to the surprise of everyone—officials included —they found quantities of oysters larger and apparently older than those on the advertised bank*. The inspection records for the preceding four years, if the examination had been efficiently carried out should have indicated the presence of oysters each year at this locality. The actual record is, however, as follows† .—

> "1895 }
> 1896 }   ...   Bare of oysters.
>
> " 1897   ...   Not examined.
>
> " 1898   ...   Oysters plentiful, 35 to a dive, 2 inches in
>   size, healthy in appearance.
>
> " 1899   ..   Nothing of value." (sic !)

The inefficiency of present inspection methods is palpable. The oysters fished in 1900 were estimated by Captain James as four years old (loc. cit.), so that by the Inspector's own showing this particular bed was missed on two occasions out of the three that it was examined. Oysters do not and cannot migrate, and if the oysters seen in 1898 and fished in 1900, were missed in 1896 and 1899, we cannot do otherwise than condemn the character of the methods employed in inspection.

Who can say how many similar oversights there have been? Careful scrutiny of the inspection records discloses many suspicious entries. Take the Karai Karuval Pār, one of the most productive banks in this region. The records for 1897—1902 show that oysters were found during the first three years, but in 1900 when they should have been ready to fish, no inspection was made and in 1901 the bank was reported bare of oysters.

Can we doubt that a fishery was missed in 1900?

The records of the Velangu Karuval Pār and Trichendūr Puntottam Pār are identical.

On the Odakarai Pār in 1899 there were "oysters, 20 to a dive 2½ inches in size, healthy. A very small quantity of dead shells

---

* "Madra Board of Revenue Proceedings, " No. 208, dated October 1900, page 4.

† Copied from the Inspection Registers in the office of the Superintendent of Pearl Fisheries, Tuticorin.

were found. Divers report that the undertow was very heavy and that they had much difficulty in keeping on their feet. Large quantity of weed on this bank," but no examination was made in 1900, 1901 or 1902!

Mr. Sullivan Thomas remarked the same discrepancies in the inspection records.* He says:—

"Looking for instances of oysters that have been obviously "missed, we find that in 1869, banks 15 and 16 contained oysters "2½ and 3 years old in December, where a blank was recorded in "March of the same year. Again, in March of the same year, bank "49 held 'many oysters of 1, 2 and 3 years of age,' which for want "of inspection had not been found before. In April 1878, banks 44 "and 45 'were thickly covered with oysters of one year age,' and in "May of the following year the record is 'blank.' If they had not "migrated and been missed, we might perhaps have found some "traces of at least a few dead shells. In 1882, we find 'dead shells' "of we know not what age on bank 20, which the previous year "was 'blank.'

"Banks 15, 16, 17 might seemingly have been fished in 1870, "but they were not inspected. Perhaps Captain Phipps was away; "perhaps the necessity for inspection was lost sight of for want of "statement B."

I may add that long weeks before I made my investigation and before I had acquaintance with the facts above related, my most intelligent coxswain, a Tuticorin man himself, in reply to my inquiry if he had any theory why oysters came to maturity so seldom on the Indian banks, said "oysters often come, inspection not good, not wide enough." He remarked that he and his people often said among themselves that if the Indian inspection was carried out in the thorough manner it is on the Ceylon side there would be more frequent fisheries. As he said, long ago fisheries were very good off Tuticorin and Kāyalpatnam,—why should they now be so very few and unprofitable? This opinion expressed, I believe, his honest belief; there was no advantage in deceiving me and at that time he had no idea that I was likely to have any connexion with the Indian banks. Candid opinion of the native fishermen is often shrewd and well considered,

* " Report on Pearl Fisheries and Chank Fisheries," Madras 1884, page 24, paragraphs 76 and 77.

and I agree cordially with Mr. Sullivan Thomas in his remark "as regards fisherfolk knowledge—it is marvellously good and should never be neglected, but at the same time always tested "* In other words the ideas of the local fishermen and divers may often furnish a valuable working hypothesis

(b) NUMERICAL DEFICIENCY OF DIVERS ATTENDING THE FISHERIES

Apart from any question of the fertility of the banks, the inadequate supply of divers attending the Tuticorin fisheries has frequently entailed disastrous financial consequences, notably in 1889 and 1890 In those years large fisheries took place concurrently off the Ceylon coast, and as the Ceylon fisheries are believed by the divers to yield them better results than those on the Indian coast, it was with considerable difficulty that any men were prevailed upon to attend the latter This state of affairs was well known among the native merchants and all the more wealthy resorted accordingly to Ceylon as the market possessed of the greater attractions Their abstention further influenced the results adversely.

Take the fishery of 1889 for example. In that year the Tola-yiram Par was densely stocked with fine oysters nearly six years old. Captain Phipps, the then Superintendent of Pearl Fisheries, calculated that there were 309,760,000 oysters upon the bank; but for want of sufficient boats and divers the gross take, 12,600,000 oysters, barely reached 4 per cent of the estimated total available. The average number of boats out per day was 35; the largest on any one occasion was but 48.

The next year, when the oysters were dying off, an even worse state of affairs prevailed; the average number of boats employed per day fell to 21 and the total take of oysters was a miserable million and three-quarters (1,806,762), bringing in a paltry profit of Rs. 7,803 to the Government

The ensuing year, as was to be expected from the age limit being exceeded, no oysters were found on the banks

The combined takes of 1889 and 1890 were under 14,500,000 oysters, so that if we accept Captain Phipps' estimate of over 309,000,000 on the bank in 1889, the Government harvested a wholly

---

* *Loc. cit*, page 25.

inadequate proportion of the crop   Can we justly characterize the
Indian banks as being poor and unsatisfactory when one bank
brings such a multitude to maturity in one year? Is it not
more reasonable to lay the blame on antiquated methods and lack
of foresight and method in organization? The average price
obtained per thousand in 1889 was Rs. 22-8-6; therefore if the
organization of the fishery had ensured, as it ought to have done,
the lifting, we will not say of the whole 309,000,000 of Captain
Phipp's estimate, but merely of a modest 50,000,000 oysters, then,
instead of taking but Rs 1,89,986 in 1889, Government would
have had a revenue of Rs. 7,50,000—a preventable loss occurred
of over 5 lakhs of rupees

The actual take was, however, as I have stated, but 4 per cent
of the estimated crop; 96 per cent was literally thrown away for
want of the means to gather it in.

To overcome the labour difficulty created by the preference
shown by the divers for Ceylon when fisheries coincide in the
same year on each side of the Gulf of Mannar, it was recommended
by the Board of Revenue in August 1890 that efforts should be
made to arrive at some arrangement with the Ceylon Government,
the basis of arrangement to be either a division of the fishing
season in point of time or a limitation of the number of boats
employed upon the Ceylon side.

Subsequently an agreement was actually arrived at[*] upon the
former basis whereby when fisheries on the two sides of the Gulf
should occur in the same season in any future year, it was agreed
that the Ceylon fishery should begin in February and close at the
end of March, leaving April and May for the prosecution of the
Tuticorin fishery.

From the experience I have had of actual fishing conditions, I
am of opinion that in practice this agreement will be found unwork-
able. The beginning of February is too early in the season to start
fishing on the Ceylon coast. Divers will not attend till weather
conditions become settled, till the intermonsoon lull begins, characterized by alternating land and sea breezes and by clear limpid
water free from suspended particles of mud and sand

No dependence can be placed upon the oncome of this period
prior to the first week in March, and I cannot see how the Ceylon

---

[*] In February 1892, according to information supplied from the Colonial Secretary's
office, Colombo.

Government can agree to close their fishery some three weeks after the date of actual opening and just when the fishing is probably at its best. Apart from governmental considerations such a proceeding would be deeply resented by the divers and the merchants ; if they were compelled to go, the fishery being summarily closed, the consequences would be felt at subsequent fisheries. The proposal is only practicable if fishing could be begun early in February and this as I have said is impossible owing to circumstances beyond the control of the Ceylon Government. Neither can the Ceylon Government limit the number of boats participating if there be sufficient abundance of pearl oysters to justify the work, indeed it would be an advantage to Madras if the Ceylon Government were able to obtain such a number of boats as would clear the bank to be fished in a limited period, as then the divers would be at liberty to depart and would be available for the Indian beds. However, even in the case of a cessation of the Ceylon fishery at the end of March, I am convinced that an Indian fishery in April and May would benefit thereby very little if the Ceylon fishery had been at all successful. At a fishery such as the Ceylon one of this year (1904), the divers make so much money as to be wealthy beyond their dreams of avarice for that year at least. The more prudent Moormen have made enough money to enable them to invest in new fishing nets, new boats, or jewellery and dowries for their women, while the thriftless Paravas find enough money in their pockets to hand a substantial sum to their Church and leave enough over to permit them to feast and be merry for several weeks or, perhaps, even months.

Such men will be induced only with the utmost difficulty to undertake a second diving season hard on the heels of the first ; they will be restless, discontented, and eager to seize any excuse to get away. Witness what happened in 1890 when a number of divers returning from fishing on the Ceylon banks, were persuaded to resume operations off Tuticorin ; only eight days fishing was obtained as the divers utilized with their usual skill the stalking-horse cry of " sharks on the banks." As any stick is good enough to beat a dog, so any excuse is considered good enough to utilize when the divers for any reason wish a fishery to come to an end. At one time it is "sharks", at another, the alleged scarcity of oysters, "chippi illei." Illness, rumours of cholera, small profits, rough weather, chill winds, are all utilized with the utmost cunning but the true reason—that they have made enough money—is always kept in the background

Hence I conclude that relief must be sought in some other manner and that it is necessary for the Madras Government to proceed entirely independently of the Ceylon authorities and to accept, as an unpalatable but none the less living reality, the fact that till present conditions be radically reformed, the Tuticorin and Kilakarai divers have not the requisite confidence in the Tinnevelly pearl fishery administration to induce them to forego attendance at a Ceylon fishery when such clashes with one on the Indian banks

Many years ago Captain Worsley, when acting as Supervisor of the Ceylon Pearl Banks, summed up his conception of the Inspector's duties towards the oysters under his charge in the dictum "find them, watch them, fish them." I have shown that the organization of the Indian Pearl Fishery Department has failed notable in all these operations, lamentably so in 1889

Detailed inspection carried out with scientific accuracy by a capable officer endowed with biological knowledge and with acquaintance with elementary marine surveying, furnishes a sufficient remedy for the first and second of these administrative diseases ; the third is more difficult to cure, though much improvement might be counted on as certain to take place when the divers become aware of the improvements taking place in the methods of inspection. With confidence in their Inspector and in the statements he might publish regarding the promising character of a bank about to be fished, many would, I believe, voluntarily remain at home in spite of Ceylonese counter attraction

This we must not, however, count upon till the new organization proves its efficiency by results, and we come back again to the problem, how can we fish a large number of oysters, say 50,000,000, during a fishing period not exceeding eight weeks (March and April), in spite of the defection of the great bulk of the local divers ?

I can think of but two alternatives, (a) the utilization of mechanical means and (b) the drafting to the fishery of a sufficient body of Arab divers

Regarding the former plan, although the character of the bottom on the Tolayiram Pār is favourable to the employment of the dredge, the numbers of oysters to be dealt with are so enormous and the occurrence of fisheries so erratic and occasionally so long deferred, that at present I cannot see that this is a practicable solution, so long as the fishery be conducted by Government   A fleet of dredging vessels would be required and the maintenance of these cannot be justified till a cultural scheme be perfected which will ensure tolerably regular periodic (annual) fisheries.   The most that is feasible is to fit the fishery steamer with dredging equipment and so enable her to do her share in the actual fishing operations *   The same equipment would serve for the dredging of young oysters for the purposes of transplantation, and it might also be utilized for the dredging of chanks, though I doubt whether the results from the last-named work would be sufficiently remunerative and would counterbalance the extra expenditure that would be occasioned in coal and oil.

The alternative of obtaining a supply of Arab divers adequate to work the fishery is left us   It appears to me that if due precautions be taken to obtain true Persian Gulf divers in small gangs under men who can give adequate monetary guarantee for the good behaviour of the men supplied by them, that this plan is eminently feasible

At the present year's Ceylon Fishery (1904) 258 Arabs were allowed employment and Mr Lewis, Superintendent of the Fishery, states in his report †

"As the fishery proceeded and the advantage of having them "had become apparent, I was prepared to take more.   They gave "very little trouble, and were very useful both for the starting of "the fishing and for keeping it going towards the end.   They were "always most keen on going out, no matter what the weather was, "and they rather roughly handled a Jaffna tindal who started for "the fishing one morning but turned back because his sail split. "They offered to mend it for him, but he said he had no materials, "Their indignation was great, and they were loud in their com-"plaints   They are as used to handling boats as they are to "diving, and had great contempt for tindals who were deterred

---

* The results obtained during the Ceylon fishery of 1905, show that an average of 55,000 oysters may be reckoned as the daily catch of a properly equipped small dredging steamer under good management.   The cost of wages and up-keep is considerably less than the value of the divers' share of oysters, so we find dredging to be a more economical mode of fishing than the employment of divers on the one-th rd share basis, provided work can be found for the steamer in the off season ; this latter is the difficulty

† " Reports on the Pearl Fishery of 1904 "   Sessional paper No. XIII, Ceylon, 1904, page 6.

"from proceeding to the banks owing to small accidents to their
' boat or gear "

Further evidence of the good work and reasonable disposition
of these Arab divers when treated justly, is afforded in Captain
James' report on the 1900 Tinnevelly fishery, * his words being--
" At first there must have been quite 1,500 divers, of which about
" 200 were Arabs These latter I consider quite the best men to
" have at a fishery, quiet, good-tempered and hardworking, and
" quite amenable to all discipline, much more so than the Parawas
" who are a constant source of trouble, both on the banks and in
" the Kottoo, where they were constantly being caught conceal-
" ing oysters, which of course were always confiscated Only one
" Arab was caught doing this, and his companions abused him for
" disgracing them The Malayali divers left the banks after the
" first few days as the water was too deep "

Fortified with such favourable opinions from men who had to
meet and control these divers ashore, where trouble is more likely
to occur than at sea, I have no hesitation in saying that I have the
highest possible opinion of these men and of the quiet, methodical,
and energetic manner in which they conduct their work I watched
them at work daily throughout the last two fisheries, and they
were ship-companions with me when toward the end of the 1904
fishery they agreed to fish from the Government steamers

Such daily contact afforded me opportunity to obtain insight
into their characters and as a result I found them more willing to
obey my orders and follow suggestions than either the Paravas or
the Kilakarai Moormen--a result due naturally to their higher
intelligence. Quick tempered they are and restive under even the
suspicion of injustice, but withal reasonable and eminently amenable to fair treatment Personally I should not hesitate to run a
fishery entirely with Arabs, and if ordinary precautions were taken
to exclude the scum of Bombay, I am satisfied that perfect order
would prevail

During the north-east monsoon, numbers of these men visit the
ports of Kanara and Malabar, whence they might readily be
obtained †

---

* " Proceedings, Board of Revenue, ' Madras, No. 208, 1900.

† Since the above was written, I have had experience of another large pearl fishery
at which a largely increased contingent of Arabs, some 2,000 in number, was employed
Their conduct was again eminently satisfactory They gave no trouble whatsoever

(c) BANKS OF GREATEST VALUE

Descending to matters of detail, the present investigation shows that certain of the pars or rather certain groups of pars are more worthy of particular attention than others The same conclusion has been drawn with regard to the Ceylon Pars , some are clearly to be classed as favourable to the maturing of oysters, while others —the majority- are wholly unreliable in this respect. *

Of the banks off the Indian coast, historical, physical, and biological evidence combine to show that the Tolayiram Par and the Kudamuttu and Karuwal groups of Pars are the highest in relative importance, bearing the more frequent spat-falls and yielding the major number of the fisheries that we are able to localize.

The northern or Kilakarai division is of little economic importance ; prolonged inspection is not requisite in this region and the time formerly devoted to this purpose can be employed to better advantage in making more detailed examination of the pars of the Central division and in carefully prospecting in the region lying between the Karuwal group and Cape Comorin

The region last named has been neglected almost entirely in the past ; during the last 45 years only a small portion of the area has received any attention on thirteen occasions, while many square miles of sea bottom have been systematically ignored in this region, which we have conclusive evidence to show formerly yielded fisheries.

The Tolayiram Par deserves the Inspector's greatest attention ; it is the sole region seen during the investigation suitable for cultural operations The bottom resembles the better parts of the Ceylon Cheval Par and like the latter premier bank is the largest among its fellows in individual area It has also a favourable record for rearing its spat to maturity in great abundance. It may not receive so many spat-falls as the Karuwal group, but from its superior extent one successful fishery here is, if it be properly exploited by a sufficiency of divers, worth several of the smaller Karuwal group fisheries

The Tolayiram Par should be mapped into blocks and each of these should be carefully studied, periodically inspected, and the results shown graphically in chart or diagram form annually

Those parts of this region which came under my personal notice
boie but small quantities of loose stony material, " cultch " as it is
technically termed, a decidedly unfavourable factor, as the oysters
need such material for the purpose of attachment. Attention
should in future be given to this detail during inspection, in order
to ascertain if this difficiency is, as I think it is, general over the
whole area. In the event of this proving to be case means should
be taken to increase the available quantity whenever an extensive
spat-fall is found to have occurred.

(d) PEARL PRODUCTION—CAUSES OF DEATH.

Pearl production by the oysters fished in 1889 on the Tolayiram
Pār, the only bank regarding which I have any data, was less rapid
than that noted during the past two years on the Ceylon Cheval
Pār On some sections of the latter, satisfactory pearl production
is found at the age of four years, valued at over Rs. 21 per 1,000 in
the case of those fished in 1903, whereas the last oysters fished on
the Tolayiram Pār were at a similar age valued at but Rs. 3-11-5
per 1,000. It was not till they attained the age of $5\frac{1}{2}$ years that they
brought in an equivalent value (Rs. 22-8 6 being the actual average
price per 1,000 at the 1889 fishery) to that of Ceylon oysters $1\frac{1}{2}$ year
younger. The latter, however, were those from the richest known
beds and there were others which at the same age—$3\frac{1}{2}$ to 4 years—
were not rich enough in pearls to give a profitable fishery Pearl
production is, however, very variable and the yield by one generation is not necessarily a criterion as to what the next may furnish,
even upon the same ground

Examination and comparison of the Tolayiram Pār oysters of
1887 90 with those of Ceylon give fairly satisfactory results in
respect to shell growth. They are not equal to the finely grown
oysters of the Cheval, but in general appearance are of a healthy
type They are nowise stunted-looking as so many of the oysters
on the Pārs more inshore are, or, as are the oysters characteristic
of the Ceylon Muttuvaratu Pār But although they are distinctly
of the Cheval Pār type, they are of slower growth and the weight
of the shells approximated closely to that of Muttuvaratu oysters
Given an abundant infection of pearl-inducing cestode parasites,
the pearl production should be profitable in quality and quantity.
This question is still one on which we are imperfectly informed ,
the life-history of the parasite is still unsolved, and till we know

the animals which lodge the adult stage, we cannot formulate any
plan for furthering the increase in numbers of those of the larval
stage, whose presence in the pearl oyster controls the production of
valuable pearls

The ratio of infection—and of consequent pearl production—
varies greatly as is to be expected consequent upon the local
abundance or otherwise of the host of the adult parasite whatever
it may be, and also upon the relative profusion or scarcity of the
oysters themselves

Time after time I have proved by the dissection of large
numbers of oysters of the same age from different beds that the
cestode infection may vary within considerable limits and as a
consequence the pearl yield is proportionately variable For
example in November 1902 samples of the same generation of $3\frac{1}{4}$
to $3\frac{1}{2}$ years old oysters were obtained from four different beds,
with valuation results as follows : —

			RS	A	P	
Periya Par Karai	...	...	13	4	0	per 1,000
South-east Cheval	...	...	10	1	0	,,
Mid east Cheval	...	...	18	3	0	,,
North-east Cheval	...	...	23	2	0	,,

In March 1887, oysters of a similar age from the Moderagam
Par gave a pearl valuation yield of but Rs. 9-14-3 per 1,000, while
other individuals of identical age from the North-west Cheval in
the same year were valued as low as Rs 6-15-0 per 1,000

Another instance of wide variation in pearl yield occurred in
the valuation of the $4\frac{1}{2}$ to $4\frac{3}{4}$ years old oysters fished this year
(1904) from the Western Cheval Three lots varied as follows . —

			RS	A.	P.	
South west Cheval	...	.	...	36	0	0 per 1,000
North-west Cheval	.	...	...	33	12	0 ,,
Mid-west Cheval	.	.	...	20	4	0 ,,

With such wide divergence in oyster value from closely adjoining areas we can never be sure of the pearl yield from a particular
bank till we solve the riddle of the pearl cestode's life-history and
are enabled to increase artificially the proportion of infected
oysters—a matter for marine biological investigation

Meanwhile it is satisfactory to know that the oysters which the
Tolayiram Par rears are of fair quality and capable of giving a
high pearl yield

I have had no opportunity to inspect a series of successive generations of oysters from any other Indian Pār The individuals seen from the Devi, Cruxian, and other inshore Pārs appear much inferior to those from the Tolayiram Pār. They are small for their reputed age, stunted in growth, and much encrusted with sponges, corals and polyzoa In general appearance they approximate to those Ceylon oysters that hail from rocky beds—from the Muttu-varatu Pār and the Mid-west and North-west Cheval.

The Tolayiram Pār is the bank by far best suited to rear healthy oysters in quantity. Unfortunately some of the characters which render it so suitable for this, expose the oysters to heavy risks from the depredations of fishes The bare level bottom, free from clefts and crannies and boulders, gives the rock-perch and trigger fish (*Vellamin* and *Kilati*) every facility to devour enormous quantities of oysters during the first year of their existence The bank swarmed with these fishes in May last and the question of the possibility of the present young oyster population coming to maturity depends largely on whether there be many more oysters present than can be consumed by these fishes in nine months or a year When about one year old the shells become stout enough to resist the sharp teeth of these fishes and the survivors have a fair chance of living the allotted span of oyster existence, if the bank be not harried by a shoal of oyster-eating rays (*Rhinoptera* spp) These fishes, the principal enemies of the adult oyster, are often of large size, five feet or more across the disc and with mouth armed with milling teeth of great crushing power They are able to feed only upon comparatively level ground and unfortunately the Tolayiram Pār is of this character On the Ceylon side, I once walked over an oyster bed ravaged at the most but a few days previously The sight was one never to be forgotten, everywhere the flat rock surfaces, originally densely packed with oysters, as evidenced by occasional clumps remaining, and by multitudes of torn byssal cables adhering still to the denuded surfaces, were stripped in large part Wide lanes had been ploughed through, every oyster gone within the breadth of the lane At frequent intervals lay piles of broken shells, crushed flat as if passed through a mill

It is a significant fact that this ground is particularly " clean," free from cultch and from any impediment to an animal scraping the oysters off in wholesale quantities. It is ideal dredging ground. Equally significant is the fact that on rougher ground and on areas where bulky cultch occurs, no depredation whatever took place. From this I infer that the presence of fragmentary material is a safeguard against rays; they are unable to differentiate between oysters and rubble when feeding, and when the latter is present, mastication being prevented, the rays find the ground unsuitable and move away.

Hence the cultching of the Tolayiram Par would serve two purposes of vital importance; it would give additional and much needed holding ground to oysters and would tend largely to diminish the damage liable to result from the inroads of rays.

Much more sediment is held in suspension in the water on the Indian banks than in the case of the Ceylon banks. I do not however consider that this exercises any greatly deleterious effects upon oysters on the outer banks of the central and southern divisions; on the Kilakarai banks the profusion of muddy sediment is excessive, as it also is on some of the inner of the more southern banks, and in such places we cannot expect any spat-fall ever to reach maturity. From the mouths of all the rivers along this coast great amounts of mud are poured forth annually and this in conjunction with the growth of new fringing coral reefs along the shore, each successive one further seaward than its predecessor, causes encroachment upon the sea. The old pars are thus brought more within the harmful influence of river sediment. The process is an exceedingly slow one and the danger to the beds appears greater on paper than it is in reality, even though we know that Korkai, the Kolkhoi of the Græco-Romans of 1,800 years ago, and the great pearling centre of that day, is now several miles inland, and its successor, Kayal, converted as well from a flourishing seaport into an inland village.

Again while the presence of so much sediment is harmful, at least to the inshore banks, it has beneficial effects upon the prosperity of the chank-beds, which flourish most vigorously wherever there is a plentiful admixture of mud with the sand, especially if there be much organic matter present, as happens off the mouths of rivers.

To this great abundance of mud is due the superior richness of the Tuticorin chank-beds over those in the neighbourhood of the Ceylon Pearl Banks, where the sand is composed largely of a coarse clean quartz-grit.

(e) CHARACTER OF THE SUPERVISION REQUIRED.

To place the entire management of the pearl banks under scientific control is the only way whereby the inspection methods can be satisfactorily reorganized and a permanent return to prosperity assured in regard to the pearl fishing industry I cannot well improve upon the words used in 1884 by the Hon'ble Mr. H. Sullivan Thomas, then First Member of the Board of Revenue, in his very valuable report to Government on this fishery, namely :—

"I think the deficiencies in the record of facts tend to show that though in Captain Phipps the Government has had an intelligent and painstaking officer, he has not been seconded by any scientific supervision anywhere, and that his active interest in his duties might have been turned to better effect if he had had from time to time the assistance of some one who had leisure and appliances for adding a scientific turn to the inquiries. It appears therefore that if the Government contemplate ever constituting a Fisheries Department, pearl fisheries should be combined with it and have the advantage of any scientific knowledge that department may have." *

Looking at the matter from a practical point of view, I do not consider that under present circumstances it would be advisable to engage a qualified expert in economic biology to devote himself solely to the care of the pearl banks, even though he be so exceptionally qualified as to be able to combine the duties of marine biologist with those at present performed by the Superintendent-Inspector of Pearl Banks.

To secure a really competent officer, a substantial salary would have to be allotted to the post and, as we know, pearl fishery work can only be carried out on the Madras coast for a maximum of five months in the year. On the other hand, the potentialities of profitable research in other directions are practically unlimited and I think that the time is now ripe, and economic fishery science sufficiently developed, to carry out the suggestion of organizing a Fisheries Department as suggested twenty years ago by the Hon'ble Mr. H. Sullivan Thomas

If this were done, and an officer appointed as Director, he might be instructed to give his primary attention to the reorganization of the pearl bank inspectional methods, the proper charting and

---

* *Loc. cit*, paragraph 96, page 29

landmarking of the beds, the elaboration of a scheme for the culture of oysters—cultching and transplantation chiefly, the recruit-ment of an adequate diving labour force prior to any fishery and, if possible, the means for the mechanical raising of oysters by means of dredges and trawls.

Control of the chank fishery should be placed with him. He would elaborate fishing methods, experimenting especially with a suitable modification of the oyster dredge, if successful, he would take steps to ensure the adoption of such improved methods by the native chank fishers. He would also investigate the feasibility of the artificial hatching and breeding of chanks—a promising depar-ture that opposes few difficulties to success.

Other shellfish of economic value are the Window-pane oyster (*Placuna placenta*) and the Edible oyster. Large quantities of the former have been fished in a landlocked bay in Ceylon and the lease of this fishery has yielded considerable sums to the revenue in the past owing to the fact that these molluscs yield abundance of seed pearls. In the Madras Presidency search should be made for beds of sufficient size to be worth exploiting.

Bêche-de-mer is an industry as yet little developed on the Indian coast and one susceptible of considerable enlargement.

In the economic investigation and control of ordinary sea and fresh-water fishing, the field for the exercise of the beneficient labours of a Fishery Department is boundless. It is not necessary here to enter on these desirable developments in detail, I will content myself with pointing out that the fish supply at many localities on the coast of the Madras Presidency might be greatly increased by the introduction of new methods, that a wide field for remunerative trawling awaits the capitalist on banks as yet scarcely touched by the native fishermen; that much help could be given to the latter by a fishery expert in teaching improved methods of net tanning and by experimenting with new fibres, such as ramie, for the production of nets cheaper and stronger and of better lasting properties than the materials now in use; that the cause of public health would be greatly served by the oversight that would be given to fish-curing yards.

A general survey of present fishery methods would be one of the results of the working of the department suggested, and from the facts ascertained it would be possible to consolidate present

20

fishery laws, modifying or enlarging the scope of such enactments as might be found advisable

The field for improving and augmenting the fish supply from fresh-water sources is still more extensive   Practically nothing is done among the natives to improve the quality and the quantity of fish in tanks, a branch of work offering immense scope for well directed cautious efforts   The restocking of inland waters that dry up annually with selected fry of species characterized by rapid growth and good table qualities should be taught, encouraged, and organized on a practical basis   Were this done, the results obtained in other countries and even in some parts of Northern India justify the prediction that the fresh-water fish supply of the Presidency would be doubled in quantity and greatly improved in quality within a very short period   Nowhere in the world are the potentialities of aquiculture greater than in India and yet nothing has been done to utilize modern piscicultural knowledge

B.—RECOMMENDATIONS

(I)

IMPROVED SYSTEM OF INSPECTION

(a) *The preparation of reliable charts* —The present charts of the Pearl Bank region are extremely unsatisfactory   The positions of none of the many landmarks dotting the whole length of the Tinnevelly and Madura coasts are shown.   It is quite impossible to lay off the ship's position with exactitude upon certain of the banks because of this deficiency, numbers of good marks—chapels, mosques, topes and the like—are in sight, but because their existence has been ignored by the cartographer they are practically useless for the purpose of the inspection of the banks, even actually misleading if we attempt to fix their positions on the coast line and fail, as is probable, in placing them correctly.   This lack of beacon indications upon the charts is further adverted to in section (d) below.

The scale of the charts in use—one mile to the half inch—is also too small for careful survey and for the insertion of the necessary details in regard to the distribution of oysters, rock, and sand in the areas inspected

All the charts used for fishery work should be on the uniform scale of I nautical mile to the inch.   A 2-inch scale is unnecessarily great and is unwieldy to handle   It is a size especially inconvenient in making comparisons of surveys effected and in furnishing comparative diagrams of oyster distribution to accompany the periodical inspection reports

In the past there has been unnecessary subdivision of the potential oyster-bearing area, resulting in the creation of 64 so-called banks. Many of these are extremely small patches of rocky bottom often not more than half a mile long by a quarter in breadth. Many again lie adjacent to one another and hence lend themselves readily to a system of grouping. I propose therefore a grouping of the banks in the manner shown upon charts I and II annexed.

Each group may be denominated by the name of the best known bank included. The grouping suggested is that which has been detailed fully in the section dealing with the topography of the banks (*vide* p. 50 *ante*) and which need not be here recapitulated.

Accompanying the revised working chart, which should be put in hand at the earliest opportunity, should be a list of at least three cross bearings taken from the central point in each inspection circle (see *infra*).

(*b*) *Adoption of a system of detailed " Circle-inspection."*—To ascertain the presence and distribution of oysters over the whole of the effective Pearl Bank region, an exhaustive examination by what I term " Circle-inspection " is absolutely essential.

Any bank found bearing oysters should be inspected by this method so long as they remain, and all hitherto unexamined ground should be covered with a network of tangent circles to ascertain the distribution of rock and sand and the potentialities of oyster-bearing.

Picked divers should be employed for the work and the services of the same men secured permanently by giving them either an annual retaining fee or an extra rate of pay. They should be placed under the charge of four inspection coxswains, also on a permanent engagement for the annual inspections in the same way as has been adopted with marked success in the Ceylon service

The banks grouped as suggested in the preceding section should next be mapped out into circular inspection areas which may be termed "Inspection-circles," of $1\frac{1}{2}$ mile in diameter, each denoted by a serial number and, where it can be done with advantage, by a distinctive name. The larger banks, like the Tolayiram and

Manappad Párs, will require several circles to cover them, whereas in the case of the smaller párs several will frequently have to be grouped within one circle.   In the latter case, the circle for convenience may take the name of the largest or most important of the included párs; in the former from its compass bearing. For example the four divisions of the Tolayiram Pár may be denominated respectively the north, the central, the south, and the southwest sections, while the circle including the Karai Karuwal and the Velangu Karuwal Párs may be termed simply the Karuwal section or circle.

During examination the inspection vessel should moor as near the centre of each section as possible, and if to one side, modify the outer boat circuits to suit this as shown in the accompanying diagram.

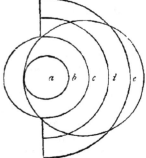

The black circle is the outline of the Pearl Bank section to be examined—
    (*a*) The ship's position.   (*d*) A ¾ mile semi-circuit
    (*b*) The ¼ mile circuit.   (*e*) A 1 mile semi circuit.
    (*c*) The ½ mile circuit

With good landmarks, reliable compass, and painstaking endeavour, it should not be difficult to anchor with approximate accuracy upon the centre of each section

The banks of superior value lie from south of Vembar to off Manappad, the Devi Pār being the most northerly, the Manapad group marking the most southerly limit.

Charts C and D * show the inspection sections which I propose. They are based upon the Pearl Bank chart at present in use and which in turn is based upon the Admiralty Chart of this part of the coast.

---

* Omitted in the present report.

Each of the circles, of which there are 35 according to my arrangement, is marked with its own distinctive number. The inspection of each circle should be completed in one morning, leaving the afternoon wherein to lift the twelve mark-buoys, shift the inspection vessel, locate the centre of the next circle, and to lay out the buoys for the following day's work.

Given average fair weather, such an inspection would occupy six weeks.

If the weather be favourable and other circumstances allow, I recommend that the whole programme be completed in one season, in which case, should the results show no considerable deposit of oysters to be present, the inspection of the following year may be greatly curtailed and be in the nature of traverse prospection rather than detailed circle inspection. Circle inspection and zigzag prospecting may be used in alternate years, but wherever oysters be found in quantity, detailed circle inspection with careful numerical estimates should be carried out annually. Where oysters of over $2\frac{1}{2}$ years of age are known to exist, inspection should take place if possible *twice* a year and a valuation sample drawn at the age of $3\frac{1}{2}$ years and thereafter at six months' intervals until such time as the valuation amounts to over Rs. 10 per 1,000, whereupon it becomes incumbent to consider whether or not a fishery should be held at as early a date as possible.

Details of the method of circle inspection.—The essential features may be stated as follows:—

Three flag-buoys are laid out by the attendant launch or tug-boat in the direction of each cardinal point of the compass at distances apart of a quarter of a mile, the inmost buoys taking their distance from the inspection vessel, which is anchored to serve as a pivot mark in the centre of the area to be inspected.

Four inspection boats (modified whale boats), each manned by a crew of six, together with three divers and two munduks, under the charge of an experienced coxswain, take up equidistant positions between the ship and the first buoy on the north line and row slowly round the ship, retaining their relative positions the while. At regular intervals the crews rest on their oars to allow the divers opportunity to make descents. The result of each dive is reported to the coxswain of the respective boat, who records it upon a diagram with which he is provided.

The four boats having each performed a complete circuit are next ranged in line abreast in the same manner as before, between the quarter and the half mile buoy and each makes a second circuit. The day's work is completed by a third and last circle, in this case between the buoys distant respectively half mile and three-fourth mile from the ship.

The four boats make a total of twelve concentric circuits, each boat making three   The results shown upon the coxswains' diagrams—each of which has three concentric circles drawn upon it representing the three circular paths covered—are transferred by the inspector to a final diagram or plan furnished with twelve concentric circles   When this has been done the distribution of old and of young oysters is graphically shown for a circular area having a diameter of a mile and a half.

After calculating in square yards the area occupied by oysters, the approximate number thereon may be estimated by taking the average number of oysters per dive (ascertained by scrutiny of the divers' results) in conjunction with the average amount of ground which a diver is credited with being able to clear at one descent. Usually this area is considered on average ground to be from two and a half to three square yards. By assuming the area per dive to be three square yards the danger of an overestimate is avoided.

(c) *Purchase or charter of an inspection depot ship*—To carry out inspection satisfactorily I recommend that either a schooner be built, purchased, or chartered, to serve as the headquarters or depôt upon which the inspection staff of divers and boatmen may live.

If purchased or built specially, the latter of which would be the more economical and satisfactory plan in the long run, cooler and more commodious quarters could be fitted up than upon a steamer, and being wooden there would be practically no liability to error in the accurate taking of compass bearings.

A steam vessel would be required for towing purposes. The "*Margarita*" might be used for the present and when it becomes necessary to replace her, the next vessel should be a screw steamer built and fitted specially for dredging and towing so that when not engaged in the latter duty, she might be used for the dredging either of chanks or of fishable oysters for market and for valuation sample

Meanwhile the "*Margarita*" should be altered and fitted to serve dredging purposes for which she is by no means unsuited.

(d) *Beacons to be charted and improved.*—An improved scheme of landmarks should be provided and the positions of the several beacons accurately fixed on the chart   It is almost incredible that none is marked on the charts in use, the Inspector has to roughly guess their relative position to the headlands and indentations of the coast indicated on the chart.   Even the Admiralty Chart, which is wonderfully accurate in other respects, shows the position of but a very few with precision—the others either being omitted or not defined with exactitude   In taking bearings from the sea, it is of little value to see upon the chart a number of marks at a certain spot indicating the presence of a conglomeration of buildings; we require the position of the most conspicuous one to be placed with precision

The beacon on Vantivu should be increased in height and an additional one erected on one of the islands to the northward

(e) *Improvements in recording the details of inspection results.*—The officer in charge of the Inspection of the Pearl Banks should be directed by the Government to insert in the records kept in his office as well as in the report furnished by him to Government at the termination of each inspection, the following details concerning the condition and abundance of the pearl oysters and associated organisms met with on each of the inspection sections, namely —

(1) The number of individual dives made upon each group of pars, and the number of those where oysters were found, together with the average number of oysters per dive over the whole of the productive ground. Not less than 300 dives should be made upon *each section*, if a reliable conception of the character and condition of the area under examination is to be arrived at The number of dives made upon the banks in the past, even upon the important Tolayiram Par, have been totally insufficient   The Tolayiram Par is of such large extent that four inspection circles are needed to cover it adequately, equivalent therefore to a total of 1,200 dives   From the office records I notice that in 1890 ninety dives were made, in 1892, 158 dives; in 1896, 220 dives— far too few to give a reliable conception of the condition of the bank as a whole.

Other banks fared even worse Taking some figures at random I find that 12 dives were made to suffice for the Alluva Par in 1886 and 32 in the following year On the Tundu Par 35 dives were

made in 1885, 31 in 1887, 7 (!) only in 1889   On the Karai Karuwal
Pār, one of the most frequently productive of the Indian banks, a
sorry seven dives sufficed for the examination of 1888, while the
Velangu Karuwal Pār had 74 dives in 1887 and 63 in 1891

(2) The average weight and dimensions of an average sample
of the living oysters found in each locality should be recorded with
exactitude   Where the oysters are numerous, the sample should
be as large as possible to diminish the possibilities of error   The
weight should be recorded in pounds, ounces, and drams, and
where possible 100 oysters should be weighed together.   In express-
ing the average weight of the individual oyster it might be useful
to express the result in grammes, as the metric system is more convenient for the purposes of comparison than avoirdupois weight.

When there are large numbers of oysters present and possibilities of an eventual fishery, the cleaned (empty) shells of 5
individuals should be averaged in like manner

I think it probable that we shall eventually find the average
weight per shell the most reliable guide in ascertaining whether
growth be satisfactory or not and also in ascertaining the approximate age of oysters of unknown history.

In the same way I recommend the dimensions to be recorded
in centimeters and millimeters, recording the length, depth, and
thickness of 25 individuals taken haphazard and without selection
from the samples brought in by the divers.

The length is the greatest horizontal distance between the
anterior and the posterior margin of the shell taken parallel with
the hinge-line, as shown upon the accompanying diagram   The
depth is the longest line that could be drawn (measured) at right
angles to the line of greatest length , it extends from the hinge to
the most ventral point of the free margin of the shell

The anterior aspect of the shell can readily be distinguished as
such because of the presence of the byssus at that side

The thickness should be measured by means of a pair of calli-
pers, clasping the jaws upon the thickest part of the oyster, a point
indicated in the diagram on the next page by the letter A.

(3) The general outward appearance, stunted or of free vigorous
growth, should be stated and also whether the oysters be extensively covered or not with sponges and other crusting organisms
in exceptional degree.

Dorsal aspect

Hinge

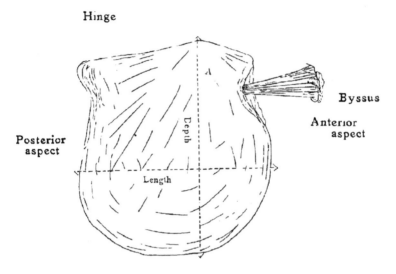

Posterior
aspect

Byssus

Anterior
aspect

Ventral margin

Diagram to explain the chief terms used in describing the shell of the pearl-oyster

(4) The comparative abundance of the following animals should be noted, so far as it is possible to ascertain the facts.—

    (*a*) Chanks (with a view to utilizing this knowledge in further exploiting the chank fishery)

    (*b*) Starfishes (especially the scarlet-lake coloured *Pentaceros lincki*, a great enemy of the pearl-oyster).

    (*c*) Rockfishes and Trigger-fishes (*Vellamin* and *Kilāti*).

    (*d*) Sûran (*Modiola barbata*)

    (*e*) False-spat (*Avicula vexillum*)

The abundance of sea-weed should also be recorded

Charting the results.—Each of the four inspection Coxswains should fill up, each day that circle inspection be employed, a simple diagram provided with three concentric circles, while the Inspector should the same day transfer to a master-form

21

provided with twelve concentric circles, the information contained in the diagrams furnished by the four Coxswains.

By this means he will be enabled to lay down the extent of the rocky bottom present, and later, when the entire inspection is complete, the outlines of these areas of rock should be filled in upon a skeleton chart. The final results, if carried out with care and accuracy would provide material, in the course of a few years' work, sufficient to enable a revision of the Pearl Bank chart to be undertaken, in respect of the pâr outlines or boundaries.

The resultant chart would then indicate the rocky areas which remain comparatively free from sand from year to year, i.e., the mean distribution or exposure of rocky bottom during normal seasons. If the distribution of oysters be also shown upon another similar skeleton chart, comparison of a series of these with the rock distribution chart would show if any part of the sandy areas frequently bear oysters, and what parts, if any, bring their oysters to maturity most regularly.

Further and much needed light would also be shed upon the relative value of different sections and would lead probably to a concentration of effort upon certain patches, while others might be found so uniformly unprofitable as to be ignored thereafter, whereby time would be economized or devoted more usefully to the more favourably situated pârs.

The Inspector, when he furnishes his periodical reports, should accompany it by the two charts named—one showing the distribution of rock and sand over the ground examined, and the other that of the distribution of oysters, a separate colour being used for different ages, the average size being given of each age.

Copies of these charts should be kept in the Inspector's office, and bound into permanent form every few years for the purpose of future reference.

(II)

REGULATIONS AFFECTING THE CAPTURE OF FISH UPON THE PEARL BANKS

Whenever a large deposit of young oysters be found on any of the pârs, if there be little sûran present, I recommend that encouragement be given to fishermen to go there and fish for Vellamin and Trigger-fish (*Kilāti*) as these are the great enemies of the pearl-oyster at this age

Stone anchors should, however, be interdicted, and the use of grapnels or iron anchors insisted on

At other times, except when the pearl-oysters are in their third year, I should recommend fishing to be permitted with the one restriction regarding the non-employment of stone anchors

When oysters on a bank approach maturity probably it would be advisable to prohibit fishing—this chiefly for two reasons, the one being the danger of disturbance of the oysters, and the other that at this time sponge-eating fish (*Holacanthus* spp), Gymnodonts, Vellamin (*Lethrinus* spp), and Trigger-fish perform a useful function in devouring and helping to keep under various competing organisms, sponges, small molluscs (sûran and brood oysters), and crusting growths that overload and overrun the valves of the older oysters

(III)

DETERMINATION OF SURFACE-DRIFT OVER THE BANKS

An accurate knowledge of the movement of the surface-water over the pearl banks is a matter of the utmost importance in their management   Without this knowledge we cannot form even an approximately accurate idea of the source whence comes the spat that from time to time replenishes one or other of our banks   So long as we are in the dark upon this subject, we cannot define in what location a reserve of oysters should be to produce the most useful results.   There are banks so situated as to be normally of no breeding value, of no importance in replenishing the banks which are our reliance ; conversely certain banks must be of supreme importance in the conservation of our beds, and it is obvious that information on these points is of vital importance in the farming of the banks   It should be ascertained whether any proportion of the spat that settles, say on the Tolayiram Pār, originated from the oyster beds on the Ceylon side of the Gulf of Mannar, whether the converse be the case, or again whether there be mutual interchange of spat

The plan offering the greatest advantages is to obtain the co-operation of the Ceylon Government in order to secure both uniformity of method and mutual assistance in carrying on this investigation I recommend that batches of small sealed bottles each containing a post card inscribed in English and Tamil, be thrown into the sea, at intervals and places yet to be determined, on both the Indian and the Ceylon side of the Gulf of Mannar, and that small rewards be given to those finders who place the cards in the hands of the nearest revenue officer or native headman, who would despatch them to the authority appointed, with particulars of the date and place of recovery.

After investigation on these lines has been carried out systematically for several years, it will become possible to determine the place of origin of much of the oyster spat, and we shall be enabled to trace the course of its wanderings while in the larval swimming condition, and in consequence know where to conserve breeding reserves of oysters for the further replenishment of the banks

(IV)

CULTURE OF THE BANKS

(a) and (b) *Transplantation and Cultching.*—The principal means whereby the banks can be permanently improved and the quantity of fishable oysters increased lies in the adoption of the correlated operations of cultching and transplantation of young oysters The latter is admittedly the most important cultural means at our disposal for increasing the harvest of the pearl banks and I am of opinion that it might be adopted with very favourable financial results on certain of the Tuticorin banks, notably upon the Tolayiram Par, provided there be proper organization of the diver labour-force, so that when the oysters become of fishable age we may be assured that the means will be adequate to bring the greater part of them ashore during the limited available season of favourable weather

If this long-standing labour difficulty be removed I advise the fitting up of the inspection steamer as an oyster dredger in order that, when young oysters are found in profusion upon unsuitable ground, a substantial proportion may be transferred to a bank where the conditions are favourable to the maturing of oysters My experience with the Ceylon dredging steamer *Violet* shows that from 500,000 to 700,000 oysters of the size attained in six months, may be transplanted during each day's employment, equivalent to a transplantation of from 15,000,000 to 20,000,000 per month— extremely satisfactory figures.

The Tolayiram Pār is a suitable bank and there I should advise
the laying of any oysters lifted from other localities, as it is in
many ways the best for this purpose. It has, however, the great
defect of possessing an insufficient quantity of loose stony fragments spread over the major part of the surface. To fit it to
receive and protect the oysters transplanted thereto and to give
satisfactory fishery results, I recommend whenever transplantation
is in operation that several hundred tons of broken coral obtain-
able from the reefs fringing the coast in many places, be spread
over the bottom where the transplanted young oysters are laid.
The cost would be comparatively small, as coral collection is a
local industry at Tuticorin and as the laden ballams and dhōneys
would proceed direct from the Hare Island reef to the bank, where
their cargoes would be scattered over culture areas marked out by
means of flag-bearing buoys

(c) *Cleaning of the Banks*—In this, as in the matter of cultching,
we may with the greatest advantage profit by the experience of
European oyster-culturists, who find it absolutely necessary to
check the growth upon the banks of all organisms other than
oysters. Not only must those that are active enemies of the oyster
(starfishes, whelks and the like), be destroyed, but also those
animals that curtail the area that oysters may occupy, and which
also consume food that would otherwise fall to the oysters. Seaweeds too are ruthlessly rooted out. As a consequence much of the
oystermen's time is taken up in cleaning the beds by means of
the dredge. If the beds are in preparation to receive spat, all
harmful matter is taken ashore—starfishes, whelks, mussels, and
the thousand and one animals that may be termed the passive
enemies of the oysters—where it finds a ready sale as manure.
Seaweeds share the same fate, while all solid material that is
overgrown with any form of life is regarded as "foul," and laid
out on the beach to be cleansed and bleached by the combined
influences of sunshine and rain

Unfortunately many of the Tuticorin banks, the Tolayiram Pār
being a notable exception, are more or less "foul." Sponges,
corals, alcyonarians, echinoderms and ascidians abound on nearly
all the inshore pārs, as for example, the Uti, Uduruvi, Kilati, and
Kudamuttu Pārs and such oysters as live there are stunted and
poor, suffering by competition with the host of creatures living
upon the same diet of microscopical organisms.

The only means to clean a bed is to dredge it thoroughly, separating and treating the materials brought up in the way above described.

The Indian banks are too extensive to permit of dredging being undertaken with this sole object in view, but, as this cleansing can and should go on concurrently with the dredging of spat for transplantation or of mature oysters for sale, we have herein one of the chief arguments in favour of taking up dredging on a scale of considerable magnitude. Sight should never be lost of the fact that dredging has four-fold utility, namely, (a) fishing oysters, (b) cleaning ground and removing enemies, (c) in thinning out overcrowded beds, and (d) spat transplantation. Its value is not properly assessed if account be taken of the first item alone or even of the first and the last.

Every live coral removed and replaced by a fragment of clean cultch may mean the addition of three oysters at the next fishery, every starfish destroyed *does* mean scores of oysters saved from destruction ; every Clione-riddled block of coral bleached on the shore will tend to reduce the widespread havoc this inconspicuous sponge causes amongst the oysters. The immense advantage that accrues from keeping the banks in a state of thorough cleanliness can well be appreciated by an agriculturist who knows how his crops fall off if weeds be allowed to run riot unchecked, if fungoid and insect pests be ignored if the soil be never disturbed and if sun and air be excluded therefrom

(d) *Thinning out of oysters.*—The evil effects attendant upon overcrowding of the oysters which so often takes place upon certain of the Indian banks have been laid stress upon, and I think sufficiently demonstrated. The remedy suggested consists of thinning out at suitable time. The dredge again is the only remedial agent. Thinning out, transplantation, and cleaning the bank may all proceed conjointly —the thinned out oysters being deposited on unoccupied ground, while the foreign organisms and the cultch materials will be taken ashore, the former to be destroyed, the latter to be bleached

(V)

CREATION OF A FISHERIES DEPARTMENT.

A Fisheries Department should be constituted under scientific control and the work of inspection of the pearl banks and superintendence of the chank fishery transferred thereto as the most important duties under its control.

Such a plan would enable these two important departments to be developed economically and on sound practical lines, would enable attention to be given to the development of other fishing industries, marine and fresh-water, at present under no scientific supervision, and finally would set free the Port Officer at Tuticorin from work foreign to the important duties involved in the charge of the port and harbour of Tuticorin, which would then receive his undivided attention

CEYLON,   JAMES HORNELL,
February 1905.   *Marine Biologist to the Govt. of Ceylon*

XI—COMMENTS ON THE RECOMMENDATIONS MADE IN 1905 IN THE LIGHT OF SUBSEQUENT EXPERIENCE.

I find nothing material requires amendment in the recommendations made in 1905, save in respect of the method of inspection and in regard to the suggestions for the improvement of the banks by culture.

Taking the latter first, experience has shown that the one all important danger to which pearl-oysters are subject, particularly in the very young and in the fully mature condition, is that of the ravages due to predaceous fish. Elsewhere * I have treated of this in some detail. Until three or four months old and even longer in the case of some fishes, the young oysters are preyed upon by almost every one of the host of bottom-feeding fishes that haunt the rocky or stony banks suitable for the settlement of oyster spat. These ravages are so widespread and extensive that unless the spat be present in enormous quantities, sufficient to satisfy the appetite of the hordes of fishes that gather to the feast, and still to leave at the end of several months, a large quantity of young oysters, no hope of an

---

* Hornell, J. An explanation of the irregularly cyclic character of the pearl fisheries of the Gulf of Mannar, *Madras Fisheries Bulletin*, Vol. VII, 1916, pp 11—22

eventual fishery can be entertained. It all centres in the race between the appetite of the fishes and the growth of the pearl-oyster's
shell to a thickness sufficient to resist the teeth of the majority of
these fishes   I can see no means of combating this danger at a price
such as the fishery administration can afford to pay. Possibly means
could be devised to kill off or frighten away these hordes of fishes,
but the cost would be prohibitive and the results uncertain at best.
Fishing on the banks (pars) off Tuticorin is already practised
to the full extent of the fishing capacity of the coast fishermen, as
they find there the best localities for bottom fishing with hook and
line. Hence unless special means be taken at a prohibitive cost,
any spat-fall that occurs must fight it own battle for existence with
its foes, unaided by the Fisheries Department   In this connexion
it must be remembered that the Madras pearl banks lie in the open
sea from 6 to 10 miles off the coast ; consequently there are long
periods during each year when all cultural operations have either
to be suspended entirely or carried on with difficulty and comparative inefficiency   The conclusion I come to, in view of these circumstances, is that none of the cultural recommendations is practical at the present time and under present conditions not so much
because they are unsound in theory, but solely on account of the
heavy cost and the uncertainty as to whether, after the expenditure
of very large sums, these endeavours would bear fruit in showing
a profit after the expenses had been provided for

With regard to the method of inspection to be followed, in practice I found the so-called circle-inspection as described in 1905, to
answer all practical purposes, it is also a method that allows the
superintending officer to supervise operations efficiently with a
minimum of trouble   the inspection ship on which he lives usually,
is the central mark of each inspection area and around it the inspection boats circle, the officer has the boats' operations under his
eye the whole time   For some purposes, especially when extreme
rather than relative accuracy be required, a system whereby the
buoys are laid out in a series of parallel lines, is preferable. The
area so marked out should not usually exceed a length of two
miles with a width between the outer lines of buoys of one mile.
In this method the inspection vessel lies at one corner of the
rectangular space to be inspected and close supervision from the
ship is more difficult than in circle inspection, the boats instead
of circling round the ship make parallel traverses backwards

and forwards between the long sides of the demarcated space
in the manner shown in the diagram below, kindly furnished by
Capt. Kirkham of the Ceylon Fisheries Department.

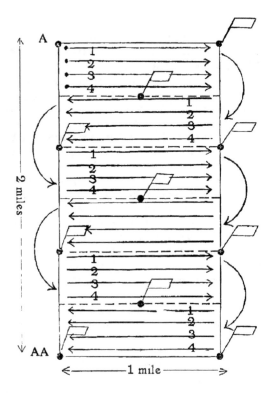

Diagram of a rectangular inspection area measuring 2 × 1 mile   A, Position of the
inspection ship at the commencement of the days work, AA, the ship's second
position to which she moves during the inspection   The flags represent the position of the 10 buoys (flagged) laid out to guide the four inspection boats in their
courses, which are indicated by red arrow lines   1, 2, 3, 4 indicate the relative position of each boat at the beginning of each traverse.

As the prevailing winds and currents are either from the north or the south, it is desirable to buoy the space to be inspected so that the long sides lie north and south,
this insures that the traverses made by the boats, rowing abreast, are equalized as
far as possible by being across the wind and current

APPENDIX A.

AN ACCOUNT OF THE CONDITION OF THE COAST OF MADURA
AND OF THE PEARL FISHERIES THERE IN THE YEAR 1663,
TRANSLATED FROM VALENTYN'S "HISTORY OF THE EAST
INDIES," VOLUME V, PAGE 146.

"1663 *Madura.*—Under the coasts of Madura are included the seven
ports or harbours, and the country of the Prince Ragonada Catta Theuver,
commonly called the ' Teuver,' with whom we are on good terms. His
contract, written on copper, is hereunto annexed But we consider a
further description necessary of the Gulf between Ceylon and Manaar, and
we shall commence with the island Ramanacoil, which is in the inner Bay,
and makes a narrow separation between the Continent of India on the one
side and Adam's Bridge on the other, but the passage, with the highest
water towards the land is scarcely six feet deep This passage is called
Pamben-aar, signifying, on account of its many windings and curlings, the
' Snake River' which gives the Theuver sufficient profit not only from the
tax laid upon the Pass, but also on account of the excursion over to the
Island, on which there stands a very old pagoda of their much revered idol
Ramana, to honour whom people come from Hindostan, Orissa and Bengal, from which circumstance it may well be imagined that a tolerable
revenue is derived

"This said Teuver is subject and tributary to the Naick of Madura ;
but since we have entered into terms of alliance and friendship with him,
he cares far less for the Naick than he did previously ; but he greatly respects us, knowing that it is in our power to take this island of Ramanacoil
from him, and therefore we can always retain him to our interests, as a
balance against the great power of the Naick, although it is not by any
means requisite that we should show any great deference to either the one
or the other

"The western lands of the Theuver are situated next to Ramanacoil,
farther eight or ten miles in the gulf, within which lie his principal places
on the sea-coast, named Wedale, Peripatnam, Killekare and Wallemoeke,
over which we have no authority But still farther to the west follow Bypaer, Bem-paer, Pattenemandoer, Tœtecoryn, Pondecail, Cailpatnam and
Manepaar, constituting the ' Seven Harbours,' all (excepting Cailpatnam,
whose inhabitants are principally Moors) being inhabited by Christian

Parruas, and provided with commodious churches. The number of these Christians consist at least of 20,000 families, maintaining themselves chiefly by diving for chanks, catching fresh fish, and diving at the pearl-fisheries when they take place, and which last employment gives them their chief profit, and causes them to live comfortably   These Parrua Christians are all under the Government of the Honourable Company, since the conquest of Tutucoryn, and they have readily submitted to our power on account of the prompt justice which we afford them.   The poorer classes are more especially well satisfied with our Government.

"Great care should always be taken to treat the people with justice and prudence, and to place a mild and sensible person as their Captain, for they are, like all Malabars, of a capricious temper and easily migrate if they are not well treated.   We have used great efforts to bring the people to our religion, but as long as the Naick of Madura, or his regent Barmi-liappa Pulle do not adopt a course different from their present one, and do not specially banish the Romish priests, we shall have little chance of attaining our object.

"*Pearl-fishery.*—The whole of the inner gulf was always under the authority of the King of Portugal, during the time of his possessing Ceylon and Tutucoryn, and on that account the Portuguese always took to themselves the full empire of the sea, including the income of the pearl-fishery which is of some consequence, particularly when diving can take place on all the banks at once, as used frequently to be the case , but for some time the banks of Manaar have given no profit, although the revenue from them was once the most considerable, and it is now fixed that they should be tried next March   But as there is some distinction with regard to the Company's interests between the banks of Tutucoryn and those of Manaar, we must give a further account of them   Whenever the pearl fishery is limited to Bempaer, Bypaer, and as far at Tutucoryn, all the oysters must be brought ashore at the last place, the market being held there and at Pondecayl, from which the Armane (as the Court of Madura is named) draws a large revenue.   The Moors are, with our permission, allowed to fish also, but they are bound to pay a large duty to the Company as may be seen in the Report of M M. Valckenberg and Boesem.

" The fishery of Tutucoryn gave last season a profit of 18,000 florins, as appears by the books of our factory at that place.   Whenever a pearl-fishery may take place at Manaar, the Company may expect much larger returns, for then the oysters will be brought ashore at Aripo, about three miles distant from Manaar, or Mantotte, being a place on the Company's own territory, and where the sale of the pearls will then be held   Y.E

should take care that a guard be stationed, to watch against irruptions of
the Wannias, Wedas, or King's people, and in order to give confidence to
the divers for themselves and their boats. If there are 100 soldiers and
100 lascoryns, the guard will be sufficiently strong, if Manaar and Jaffna-
patnam have their garrisons also strengthened."

(From the Memoir left by Governor van Goens for the guidance of his successor
Governor Hustaart, 26th December 1663.)

APPENDIX B.

Detailed Financial Statement of the Ceylon Pearl Fishery of 1694.

(Translation from the Dutch official record.)

Free Stones of the Naick of Madura, The Theuver, The Pattangatyns of Jaffnapatam, Manaar and Tutucoryn, according to old Customs.

96½ different free stones of the Naick of Madura in six boats, viz. :—

		RDS.	RDS.
4 Christian stones, at 6½ Rds. each ...		26	
1 Gentoo stone		9	
91¼ Moorish stones at 11⅓ Rds. each ...		1,052¼	
			1,087¼
60 Moorish stones in three boats for the Theuver (of which 1 is given to the Maniagaar of Pambenaar) at 11½ Rixdollars each		690	
			690
9 stones to Pariboe-neyna, Head-Moorman, viz. :—			
7 Christian stones at 6½ Rds.		45½	
2 Moorish do. at 11½ ,, ...		23	
			68½

9
—

The following to the Pattangatyns of Jaffna and Manaar :

43 stones, namely :—

10 for the Pattangatyn Moor of the Parruas of Manaar, Jan de Cruz.

10 to Anthony de Melho, Head Pattangatyn of Carreas.

RDS.

8 to the Nayenkarreas, namely :–

2 at St Pedro.

2 at Pesale

1 at Tellemanaar.

2 at Ikelampalle or Lugaar.

1 at Aripo.

——

8

——

3 to the Pattangatyn of Jaffnapatam, Don Rodrigo.

3 to the Pattangatyns of the Carreas.

3 to the Pallewellys

3 to 3 General Pattangatyns of Parruas, Carreas and Pallewellys.

1 to the Head Moorman of Jaffnapatam.

2 to the Pattangatyns' Canacapulles.

——

43 at $6\frac{1}{2}$ Rds. each ... . $279\frac{1}{2}$

——

181 stones to the Pattangatyn of Tutucoryn, viz. :—

130 to Pattangatyns, 1 Canacapulle and the Topas Moorman, each 26 stones.

6 to 2 Head Pattangatyns.

3 to 3 Canacapulles of the Community and of the Pattangatyn Moorman

1 to the Bazaar Guard.

1 to the executioner.

40 to 40 common Pattangatyns of the seven harbours

——

181 stones at $6\frac{1}{2}$ Rds. each .. ... $1,176\frac{1}{2}$

——

Total $389\frac{1}{4}$ stones valued at Rds. . ... $3,301\frac{3}{4}$

FINANCIAL ACCOUNT OF THE PEARL FISHERY OF 1694.

DR.

	FLORINS.
Expenses for the inspection of the pearl banks at various times from the year 1667 (2nd Fishery) to 1694 (3rd Fishery) as appears by the books kept at Manaar ...	6,767 16 5
To expenses incurred for the same purpose at Tutucoryn, as appears by the commercial books of that place and of Manaar ..	2,493 1 11
	9,260 18 0

Expenses at this fishery, namely, expense of soldiers and sailors, lascoreens, cooly hire, arrack, medicine, and sundry expenses of the Commissioners according to their separate account ... .   RXDS   1,709 = 5,217 13 0

Cost of 389½ free stones, which according to old custom are not paid for — see separate account ... . 3,301¾ = 9,905 5 0

To the Shroffs who counted and sorted the money and prepared it for being paid over.   40 = 120 0 0

+ Clear profit 63,057 13 0

Total ...   florins   87,561 9 0

CR.

The amount of the rent of the "change," of the bazaar, and of the clothes shop, viz. :—

	RIXDOLLARS.
The "change"	1,500
The Bazaar	137
The Clothes	31
	1,668

Stones purchased in the fishery and paid for, namely :—

1,690 Christian stones paying 6½ Rds each ..	10,985
204 Gentoo	do	do 9	do.	... 1,836
1,268 Moorish	do	do, 11½	do	... 14,242¾
				27,063¾

	RXDS.	RIXDOLLARS.
Customs formerly collected by the 'Topas-Moor of Tutucoryn, but now taken by the Company and held at the disposal of the Governor and Council of Colombo		
201 Gentoo stones at 1 fanam each ...	$20\frac{1}{10}$	
1,014 Moorish do 2 do.	$202\frac{1}{5}$	
		$222\frac{9}{10}$
12 ounces, 17 angles, and 3 as of pearls gained from oysters brought up by the divers of the Company as Wally, sold for	180	
Sifting the sand where the pearls (oysters) were laid	25	
		205
$13\frac{3}{4}$ ammonams of concealed arrack found in the bushes and out-of-the-way places at 6 Rds the ammuna	$82\frac{1}{2}$	
Deduct two-thirds given to the discoverers of the arrack	55	
		$27\frac{1}{2}$
Total . Rds		$29,187\frac{9}{00}$

or fl $87,561$ 9 0

Thus drawn up in the Fishery to the S of Aripo, 7th May 1694,

(Signed) FLORIS BLOM,
(,,) A. BERGAIGNE,
(,,) D. DE CHAVONNES,

APPENDIX C.

INSTRUCTIONS GIVEN IN 1722 BY THE GOVERNOR OF CEYLON
DEFINING THE RESPECTIVE RIGHTS OF THE DUTCH, THE
NAYAK, AND THE SETUPATI AT PEARL FISHERIES HELD
IN THE GULF OF MANNAR.

*Extracts from a despatch, dated 20th January 1722, from the Extra-
ordinary Councillor and Governor of Ceylon, M. J. A. Rumpf, and
his Council in Colombo, addressed to the Senior Merchant and
Chief Authority at Jaffna, Jacob de Jong, and his Council there* *

"It is now upwards of 22 years since the Company has indulged its
own subjects or strangers with any fishery in the Bay of Condatchy.

"The Valy, or general fishing on the Company's account, is together
with the payment for the stones, a double token of the Company's sovereignty
over the divers and the Banks from Cape Comorin, north, to Negombo,
south, or at least by these tributes enough is done to show the dominion
conceded to the Company over those seas, the bay and the pearl-banks lying
there, and the result of the enquiry of the Commissioners for the last three
years proves that this claim is indisputably made with greater foundation
than that of the Naick to the ships along the coast of Madura, when that
prince, to show his mixed authority, sets up his flag next to the Company's
standard at the fort of Tutucoryn, assists in laying down rules for the
fishery, exercises magistracy over the black people who come to that fishery,
permits all misdeeds, except treason, to go unpunished among his own
subjects during the time of fishing, and the Company winks at this and
receives tax from all pearls carried away from Tutucoryn but with the
exception of 96½ free stones he has no part or share in the produce of the
stones sold at the Banks of Madura or Aripo, which payments are received
and kept solely for the Company as Lords of these seas and bays ; but at
the same time (though it appears rather unreasonable) from old custom, a
kind of authority is exercised by the Naick over the Champanothy of every
nation, which obliges them to give to this Prince of Madura one day's
fishing free of payment, but His Highness, through his ambassadors who
came to the fishery of Condatchy, has now and then endeavoured and more

---

* Ceylon Literary Register, Volume III, pp. 166, 167.

23

especially in the year 1695, according to the custom of all black people, to institute a claim to enjoy the same tribute from all dhonies, but this has always been boldly refused to him, except with regard to his own subjects from whom he takes this tribute, as the Theuver does from his own subjects, but no further, as the Commissioners will find fully explained in the reports of 1694 and 1695, where the Company's absolute and undivided authority, if not along the coasts of Madura, at least in the Bay of Condatchy as being sovereigns there, in the same way as this is given to Princes on the coast of Madura, etc., etc., etc.

"The Maniagaars of the Armane and Theuver, as envoys sent to take care of their masters' interests in the Fishery of Aripo, must be treated with politeness and cordiality. The olas which they usually bring with them must be received and forwarded to me, and nothing must be granted to them except what is authorized by old custom, viz , to the Naick 96½, and to the Theuver 60, free Moorish stones, as appears by the lists which I mentioned to you, although the latter Prince, being limited to three boats, was accustomed to have an unequal number of stones in them, which gave rise to frequent disputes ; until at last in the year 1694 it was stipulated that whether the boats were large or small no more than 60 stones were to be employed in them, which you will unreservedly take notice of , and if any claim be made, you will refer to this rule laid down in 1694 and followed till 1699. *And as to his request of 27 free stones for the Pagola of Ramanacoil, His Excellency may give as much as he pleases from those 60 stones which are granted to him from the Valy which he receives from his own subjects,* but the pretensions of Peria Tamby, or whoever now fills his place as the Theuver's Marcair, seem better founded This claim is not a rule, but an act of liberality on the part of the Company, and granted or not, in proportion to the care and favour which he gives to the Company's trade at Kilikerry," etc., etc.

APPENDIX D

1744 ADVANTAGES TO BE GAINED BY GOVERNMENT BY RENT-
     ING OUT THE PEARL FISHERIES OF THE GULF OF
     MANNAR

*Respectful considerations relating to the renting out of the Aripo Pearl
     Banks and the Chank Fishery on the shores of the North of Adam's
     Bridge, submitted to His Excellency Gustavus William, Baron Van
     Imhoff, Governor-General, and the Members of the Council of
     Netherlands India, by Julius Valentyn Stein Van Gollenesse,
     Governor of Ceylon* (Ceylon Literary Register, Volume III,
     page 181)

Although the undersigned has not as yet acquired sufficient experience
to be able to judge fully of all matters relating to the Pearl Fishery, yet he
is unwilling to defer obeying the order conveyed in your letter of
5th November 1743,[*] and which desires that he should lay before you his
humble opinion with regard to the Chank Fishery, and also state whether
it would not be as advisable, or even preferable to rent out the Aripo Pearl
Banks, as to continue the present custom of settling whole or half fisheries,
and he hopes that Your Excellency will look over any errors in his views
of the subject, and kindly supply any defects in this statement of his
opinion

In the first place then, I must admit, as a matter beyond dispute, the
remark which Your Excellency makes in the memoir left here for the
guidance of your successor in this Government, namely that the Honourable Company is rather a loser than a gainer in our Pearl Fisheries, no
person will deny this who has a grain of local knowledge respecting the
affairs of Ceylon It is therefore necessary to seek some mode of conducting these fisheries, which may secure to the Company the profit to be
derived from them without its being accompanied by the many drawbacks
detailed in your memoir , and who can doubt that this may best be effected
by renting them out, or by selling the freedom of diving on the banks, with
a limited number of boats and persons in the same manner as now takes
place with regard to the Chank Fishery It is evident that this may be

---

[*] Merely calling his attention to the preceding remarks of Baron Van Imhoff.

done without any hindrance, and more profit will result than is expected, at all events the gain will be real and not merely ostensible It is not to be denied that at first sight some difficulties appear to rise in opposition to this plan, but the undersigned will now relate everything that to the best of his knowledge can offer hindrance, and show how in his opinion every obstacle may at once be removed.

1st Objection —The Theuver and the Naick of Madura having had from all times three days free diving in each fishery will not allow this privilege to be taken from them.

1st Answer (a).—This privilege seems to have been merely conceded because the greater number of the dhonies and people required at a public fishery come out of their country, and these will not be required if the diving takes place with a limited number of persons , the right may therefore be withdrawn

(b) If they venture to pretend that their right rests upon a better ground, and cannot therefore so easily be withdrawn, it is certain that on the other side they have never fulfilled that portion of their concessions which are laid down in legal contracts between the Company and themselves and the Company is therefore fully authorized to deny their right, even if it can be called by that name.

(c) If there remain any doubt that this can justly be done, yet this need not prevent the rentings, as their privileges may still be guaranteed to them under proper restrictions

2nd Objection —It will be difficult to find persons of so much property as to pay the price of the rent in advance.

2nd Answer —Even if they be not found in this island, speculators enough will come from the coast, and even money enough exists among the Ceylon merchants, for many together will make a Company to take shares in the adventure.

3rd Objection —Even though the number of the dhonies be limited speculators will arrive from all sides, and there will be as large a crowd of persons to purchase the pearls as ever there was at an open fishery, and then the Company will not obtain its purpose in this respect

3rd Answer (a) —It is very different from an open fishery which is proclaimed on all sides, and to which all persons are invited, but in a rented fishery it would only be necessary to give orders that no person should be admitted except those who are absolutely required to be present, and the uninvited might be sent away

(b) The oysters might be opened on the shore by the renter's people, and might be taken away at pleasure, but if it be imagined that this would bring too great a concourse of people to this island, it would be easy to order the renter to take away the oysters with him to the coast, as is done with the Chanks, and not to allow him to land them on this side the water.

4th Objection—For a complete fishery 800 or 1,000 boats are required, and how could then the work be done with a limited number of 25 or 50

4th Answer (*a*)—In the memoir already quoted a full and complete fishery is excepted from being rented

(*b*) But the same rule might hold good even in a full fishery, for (1) as a fishery seldom lasts longer than 24 days, a rented fishery might last three times as long; (2) the bank which could not be open in one year, might be rented the following years, as the assertion of the Commissioners at the last fishery seems very improbable, that the oysters being too mature loosen the pearl and let it drop; this may be the case with some few of too full a growth, the place of which others will supply which were not so mature previously

5th Objection.—It has just been answered to an objection, that what cannot be done in one year, may be done in the next one or two years immediately following, but since it has happened that there have been full fisheries for many successive years, how is it possible that these continued full fisheries could be carried on with a small number of dhonies? and then the loss to the company could be exceedingly great.

5th Answer (*a*).—A moderate profit in a rented fishery would be far more advantageous to the company than great *apparent* gain in an open fishery, at which if all matters would be weighed and balanced, the company really gains nothing; (*b*) it has not yet been proved that the oysters lose their pearls so quickly, and it is therefore uncertain if the company would sustain any injury by the delay.

6th Objection—The renter will fish the banks so bare, that the profit of the company will be quite ruined

6th Answer (*a*).—I cannot perceive why a small number of divers should strip the banks more than a greater number

(*b*) If that idea should prove to be well founded, proper directions should be established on the subject, and it must be prohibited to bring up small or young oysters; and although it is desirable to get rid of the trouble of having constant guard over the banks, yet it would not be very difficult to have two or three persons commissioned to see what goes on

7th Objection.—It will be necessary to inspect the banks in the same manner as previously, in order to know how the conditions of the rent are to be made out, for certainly speculators will make large or small offers according to the greater or fewer appearances of profit, and there will

24

always be differences of opinion; for the renter will constantly urge that the duty was not well performed, in order to obtain some deduction for his amount of rent.

7th Answer (*a*).—The renter may have full liberty to obtain indemnification from the native inspectors, in the event of an incorrect report being given in by them

(*b*) In the conditions care may be taken to guard against all after-claims, and to let the banks in whatever condition they may be found

(*c*) Public notice may be given that persons inclined to make an offer for the banks may be present at the inspection of them

8th Objection.—This rent will prejudice the chank fishery, for this latter will be at a standstill from the want of divers.

8th Answer—If divers can be found for 800 or 1,000 dhonies, then it can surely not be thought that they will be so scarce as not to be found for 50 boats, and both fisheries may easily go on at the same time.

9th Objection.—The inspection of the pearl banks takes place in November, and it is late in December before the Government is able to make out the conditions of the fishery which is to be held in the middle of February Now, it would be impossible to fix a day for offering the rent before the beginning of February, in order that speculators from the coast may have time to come to Ceylon If then there should chance to be no speculators, or if they should not make an offer large enough, it would be too late to commence preparations for an open fishery, and Government would be compelled to be satisfied with a bidding however small, lest it should be deprived of the advantage of fishery on its own account or of letting out the fishery.

9th Answer (*a*).—It is not one instant to be doubted, but there will be a sufficient number of bidders; (*b*) at all events, even if they should bid little, and we should be compelled to accept their trifling offer, it would always be satisfactory to think that the gain is clear profit

Finally, with respect to the Manaar and Calpentyn chank fishery, the Governor is of opinion that a diving on the company's own account would be far more profitable than renting the fishery, and if we resolved in our sitting on the 11th January last to rent that fishery again, it was because Mr. Raket, the Chief Officer at Manaar, was so indifferent upon the subject as to hold out no hopes of a successful attempt on our own account, yet we have since then given up renting that fishery, and it now takes place at our own risk, and although it is as yet not by any means so well conducted as it ought to be, still we by no means doubt but the company will derive a larger profit from it than 4,800 Rixdollars, which were offered for the rent of the last fishery

The undersigned hopes he has now satisfactorily obeyed Your Excellencies' wish, and clearly proved first, that the renting of the pearl banks is in every respect preferable to having an open fishery, second, that it would be better to dive for the Chanks at Calpentyn and Manaar on the company's own account.

Note—Neither the Governor General nor Governor Van Gollenesse appears to have had any actual experience in the conduct of a pearl-fishery. The latter's arguments are special pleading, apparently set forth with a view to curry favour with the Governor-General whose predilection towards leasing out the fishery was on record. J. H.

APPENDIX E

MARKS FOR THE RĀMNĀD AND TINNEVELLY PEARL BANKS ACCORDING TO CAPTAIN WICKS (1885).

Serial No	Name	Depth in fathoms	Marks and bearings.
1	Pamban Karai Par ..	$4\frac{1}{2}$ to 5	....
2	Velangu Par	$4\frac{3}{4}$ to $6\frac{3}{4}$	Roman Catholic beacon with cross on Kundagal Point is in line with lighthouse.
3	Musal Tivu Par ...	$4\frac{3}{4}$ to 6	Manamalli beacon, N. by W (westerly); centre of Musal Tivu, W. by N $\frac{1}{2}$ N.
4	Cholava Karai Par ..	$4\frac{3}{4}$ to 6	Lighthouse, N. E $\frac{1}{4}$ E Manamalli beacon, N W $\frac{3}{4}$ W
5	Kilakarai Vellai Malai Velangu Par	7	Valinukam Point, N. W by W. (westerly) Sheeramudali island, N. N. E $\frac{1}{2}$ E
6	Vellai Malai Karai Par ...	6 to 7	Sheeramudali Tivu, E by N. $\frac{3}{4}$ N ; Vellai Malai Tivu, N W.
7	Anna Par	$5\frac{1}{4}$	Valinukam beacon, N W by N , Anna Par Tivu, N. $\frac{1}{2}$ W
8	Nallatanni Tivu Par	6 to 7	Large tree in line with centre of Nallatanni Tivu, W N.W., 4 miles distant ; Valinukam Point, N by E $\frac{1}{2}$ E., $4\frac{1}{4}$ miles distant.
9	Upputanni Tivu Par	$5\frac{1}{2}$ to $5\frac{3}{4}$	Mukur church, bearing N W. by N $\frac{1}{2}$ N. Mukur church is in line with the north end of Upputani Tivu, and large tree on Nallatani Tivu, N E.
10	Kumulam Par	6	Church at Mukur is in line with the middle of Upputanni Tivu.
10 A	Valinukam Par	5 to 6	Valinukam Point, N $\frac{1}{4}$ E. $1\frac{1}{2}$ miles, centre of Anna Par Tivu, N.E by E $\frac{1}{2}$ E., 3 miles.
10-B	Valinukam Tundu Par .	7	Valinukam beacon, bearing N by W. $\frac{1}{2}$ W.
11	Vembar Periya Par ..	9 to $10\frac{1}{2}$	Mukur church is in line with centre of Upputanni Tivu
11-A	Karai Par	6	Church at Mukur is just open to the southern extreme of Upputanni Tivu.
12	Vaippar Periya Par	6 to 7	...
13	Karai Par . ..	$6\frac{1}{4}$	Volkart's chimney open to south of low Valnad hill ; Hare Island lighthouse, S. W. by S. $\frac{1}{4}$ S Five miles south from the mouth of the Vaippar river. Putnamurdur trees bear W $\frac{1}{4}$ N. $6\frac{1}{4}$ miles
14	Devi Par . .	6 to $7\frac{1}{4}$	Volkart's chimney in line with southern end of low Valnad hill. Vaippar large tree just over north end of Chulli Tivu
15	Pernandu Par ..	6 to $6\frac{1}{4}$	Tuticorin port flagstaff in line with south end of small Valnad hill. Putnamurdur trees, W by N. $\frac{1}{2}$ N. with round hill just open to north. Volkart's chimney between two Valnad hills, also for the Pernandu Par.
16	Padutta Marikan Par		
17	Padutta Marikan Tundu Par.	6 to $6\frac{1}{4}$	Volkart's chimney in line with centre of large red Valnad hill. Putnamurdur trees is well open to the Black hill.

Marks for the Râmnâd and Tinnevelly Pearl Banks according to Captain Wicks (1885)—cont.

Serial No.	Name.	Depth in fathoms.	Marks and bearings.
18	Tuticorin Kuda Par ...	7 to 7¼	Volkart's chimney under peak of large red Valnad hill, also French church the same. Black round hill in line with north end of tope, south of Putnamurdur.
19	Cruxian Par	5¾ to 6¾	Roman Catholic church under peak of large Valnad hill. Round hill open north of Kaswar Tivu.
20	Cruxian Tundu Par ..	6 to 6¼	Goa church in line with north end of large red Valnad hill, also Tuticorin cotton press chimney shut in with hill. Putnamurdur trees just inside of north end of Kaswar Tivu.
21	Vantivu Arupagam Par ..	5¼ to 6¾	South end of long red Valnad hill just clear of north end of Hare island. Top of large red Valnad hill just south of south end of trees at the back of Tuticorin. Putnamurdur trees just in line with north end of Vantivu.
22	Nagara Par	6¾ to 7	Lighthouse under peak of large Valnad hill and tree on top of red hill. Ulagamalai in line with church on Vantivu.
23	Utti Par	7 to 7½	Lighthouse in line with the north end of large red Valnad hill. Ulagamalai clear of south end of Vantivu.
24	Uduruvi Par ..	7 to 8¼	Lighthouse W. ¼ S. distant.
25	Kilatti Par .. .	7¼	Flagstaff in line with south end of Melur casuarina tope. Large peak of Valnad hill on centre of opening between Hare and Long islands.
26	Attuvai Arupagam Par .	7	Centre of Melur casuarina tope in line with bank house and northern end of Hare island.
27	Attombadu Par	7¾ to 9	Three round trees on Long island in line with northern end of Valnad hill and Salt Office in line with southern portion of Melur tope.
28	Pasi Par .	8½ to 9	Extreme northern end of Hare island, bank house and the northern portion of Melur tope in line. Three round trees on Long island in line with gap between Valnad hills.
	Pattarai Par	7½ to 8½	Protestant church in line with Hare island.
29	Kutadiar Par ...	8	Port flagstaff and lighthouse in line bearing N. 70° W.
29 A	New Bank	9¾ and 10	Position latitude 8° 51′ 30″ N. by observations, distant N. E. by E. ½ E., 5¾ miles from Nagara Par and 3 miles nearly north of the Tolayiram Par. Marks—lighthouse on Hare island in line with the S. knob of low Valnad hill. Putnamurdur trees open to south of round Black hill and behind.

Marks for the Rāmnād and Tinnevelly Pearl Banks according to Captain Wicks (1885)—cont

Serial No.	Name	Depth in fathoms	Marks and bearings
30	Tolayiram Par	8 to 11	Latitude 8° 45′ 30″ N , longitude 78° 23 00″ E Hare island lighthouse, N. 79° W , distant 9 miles.
31	Vada Ombadu Par	8 to 8½	Black hill in line with lighthouse Peak of large Valnad hill, bearing W. Volkart's chimney open short distance to south of lighthouse
32	Saith Ombadu Par	8 to 8½	Goa Mission Church in line with the middle tree of the three on the north of Long island
	Puli Pundu Par and Kanna Puli Pundu Par.	8 to 9	Palaikayal church is open to the southern extreme of the low southern Valnad hill, bearing W ½ S
33	Alluva Par	6	Southern boundary pillar in line with the centre of low Valnad hill Moravichulli Tivu beacon, bearing N 84° W. Lighthouse on Hare island, N 31° W
34	Kanna Tivu Arupagam Par.	5¼ to 6	Volkart's chimney is in line with the middle tree of the three on the north of Punnayaddi Tivu or Long island Southern boundary pillar in line with centre of the low Valnad hill Moravichulli Tivu beacon, bearing W
35	Tundu Par	...	..
36	Nenjunchan Par Par Kudanjan Par Mela Ombadu Par	8 ⎫ 7¾ to 8½ ⎬ 8¼ to 8½ ⎭	Volkart's chimney is in line with the coconut tree on Hare island A small peak in high mountains seen through gap between Valnad hills
37	Pinnakayal Seltan Par	8¼	Palaikayal church is just open to the north of low Valnad hill Pinnakayal church bearing W (southerly) Trichendur pagoda, S 39° W
38	Sandamaram Piditta Par	9 to 9½	Tiruchendur pagoda S 42° W. Kayalpattanam point, S 79° W Palaikayal church is just open to the north end of large Valnad hill
39	Irativu Kudamuttu Par	7½ to 8½	South Roman Catholic church at Pinnakayal is in line with south end of low Valnad hill
	Nadukudamuttu Par	7½ to 8½	Southern portion of tope at Pinnakayal is in line with south end of low Valnad hill
40	Kovilpiditta Pudu Par	9½ to 10	South end of small Valnad and ruined church at Pinnakayal in line
	Sankuraiya Pattu Par	10	North end of small Valnad and Agastiar Malai in line
	Nilankallu Par	9 to 9½	South end of small Valnad and Agastiar Malai in line.
	Sethu Kuraiya Pattu Par	10	... ..

According to Gautain Baker

Marks for the Rámnad and Tinnevelly Pearl Banks
according to Captain Wicks (1885)—cont.

Serial No	Name	Depth in fathoms	Marks and Bearings.
41	Rajavukku Sippi Sothicha Par.	...	North Roman Catholic church at Pinnakayal is in line with the south end of low Valnad hill. South end of Nadumalai in line with the Sea Customs Office at Kayalpattanam.
	Kudamuttu Par	7½ to 8½	...
	Saith Kudamuttu Par	7½ to 8½	Pinnakayal tope in centre of two Valnads.—*According to Captain Baker.*
42	Naduvu Malai Piditta Par	8 to 8½	Church at Kayalpattanam is open to the south end of low Valnad hill.
	Periya Malai Piditta Par	8½ to 10	...
43	Kadian Par	7 to 7½	Trichendur pagoda, S. W by W. distant 7¼ miles. Mosque at Kayalpattanam, W ¼ N , 6 miles.
	Kanava Par	7¼ to 8	...
	Pudu Par ...	7½ to 8	Armuganeri is just clear of Kayalpattanam point
44	Karai Karuwal Par	8½	Mosque at Kayalpattanam in line with extreme end of low Valnad hill. Trichendur Pagoda bearing W ¼ S , distant 7¾ miles.
45	Velangu Karuwal Par	8½	Mosque at Kayalpattanam bearing N W by W ¼ W., distant about 8½ miles and in line with the centre of low Valnad hill. Trichendur Pagoda, bearing W. 7¼ miles.
46	Tundu Par ...	9 to 9½	Trichendur Pagoda bearing W ½ N.
47	Trichendur Punthottam Par.	7½ to 8½	Trichendur Pagoda, W. by N ½ N. (northerly)
48	Odakarai Par	9 to 11	Marks and Bearings taken from " Station ship." Trichendur Pagoda, N 84' W., and in line with square-topped peak of highland behind Manappad Point S. 58° W.
49	Chodi Par	8 to 9	Mosque at Kayalpattanam is in line with centre of large Valnad hill. Trichendur Pagoda, W. by S. ¼ S., 3 miles
50	Sandamacoil Piditta Par	8½ to 9	...
51	Teradi Piditta Par	8¼ to 9	...
52	Semnian Patt Par ...	8 to 9	Manappad Point N. 37° W. Trichendur Pagoda N 2° W.
53	Surukku Ombadu Par	8½	Church on Manappad Point is in line with centre of extreme southern hill north of Cape Comorin and known as Sirumalai Trichendur Pagoda in line with the southern end of low Valnad hill.
54	Manappad Periya Par	5¾ to 7	...

SKETCH PLANS.

Nos I & II.--Sketch plans of the Central Pearl Bank region showing graphically the groups which I propose to form by the linking together of adjacent and related pârs, are appended.

Ingram Content Group UK Ltd.
Milton Keynes UK
UKHW020105090323
418239UK00006B/588